A CONCISE INTRODUCTION TO DATA STRUCTURES USING JAVA

CHAPMAN & HALL/CRC
TEXTBOOKS IN COMPUTING

Series Editors

John Impagliazzo
Professor Emeritus, Hofstra University

Andrew McGettrick
Department of Computer
and Information Sciences
University of Strathclyde

Aims and Scope

This series covers traditional areas of computing, as well as related technical areas, such as software engineering, artificial intelligence, computer engineering, information systems, and information technology. The series will accommodate textbooks for undergraduate and graduate students, generally adhering to worldwide curriculum standards from professional societies. The editors wish to encourage new and imaginative ideas and proposals, and are keen to help and encourage new authors. The editors welcome proposals that: provide groundbreaking and imaginative perspectives on aspects of computing; present topics in a new and exciting context; open up opportunities for emerging areas, such as multi-media, security, and mobile systems; capture new developments and applications in emerging fields of computing; and address topics that provide support for computing, such as mathematics, statistics, life and physical sciences, and business.

Published Titles

Pascal Hitzler, Markus Krötzsch, and Sebastian Rudolph, Foundations of Semantic Web Technologies

Uvais Qidwai and C.H. Chen, Digital Image Processing: An Algorithmic Approach with MATLAB®

Henrik Bærbak Christensen, Flexible, Reliable Software: Using Patterns and Agile Development

John S. Conery, Explorations in Computing: An Introduction to Computer Science

Lisa C. Kaczmarczyk, Computers and Society: Computing for Good

Mark J. Johnson, A Concise Introduction to Programming in Python

Paul Anderson, Web 2.0 and Beyond: Principles and Technologies

Henry M. Walker, The Tao of Computing, Second Edition

Mark C. Lewis, Introduction to the Art of Programming Using Scala

Ted Herman, A Functional Start to Computing with Python

Mark J. Johnson, A Concise Introduction to Data Structures using Java

CHAPMAN & HALL/CRC
TEXTBOOKS IN COMPUTING

A CONCISE INTRODUCTION TO DATA STRUCTURES USING JAVA

Mark J. Johnson

Central College
Pella, Iowa

CRC Press
Taylor & Francis Group
Boca Raton London New York

CRC Press is an imprint of the
Taylor & Francis Group, an **informa** business

A CHAPMAN & HALL BOOK

CRC Press
Taylor & Francis Group
6000 Broken Sound Parkway NW, Suite 300
Boca Raton, FL 33487-2742

First issued in hardback 2017

© 2014 by Taylor & Francis Group, LLC
CRC Press is an imprint of Taylor & Francis Group, an Informa business

No claim to original U.S. Government works

ISBN 13: 978-1-138-40220-1 (hbk)
ISBN 13: 978-1-4665-8989-6 (pbk)

Visit the Taylor & Francis Web site at
http://www.taylorandfrancis.com

and the CRC Press Web site at
http://www.crcpress.com

To Lyn

Contents

Code Listings

Tables

Preface

Welcome!

This text presents an introduction to data structures in Java™. It assumes some prior programming experience but not necessarily in Java.

Data structures are the building blocks used to create any significant piece of software. Just as smaller programs are made with functions, loops, and if-statements, large-scale programs use stacks, queues, lists, and maps. Becoming familiar with how to use, design, implement, and analyze these structures is an important step on the path to becoming a skilled software developer.

One piece of advice for learning this material: do as many of the exercises as you can. One of the nice things about writing programs is that you can run them to see if your solutions work or not. That quick feedback is invaluable.

Enjoy.

To Instructors

This text is designed to support a second course in computer science with an emphasis on elementary data structures. It takes a developmental approach to enhance student learning:

- More code and guidance are provided at the beginning, allowing students to adapt to Java and begin learning data structures. As students become familiar with the landscape and more independent, less code is given and more algorithms are outlined in pseudocode.

- Important concepts are introduced gradually and revisited with increasing depth. For example, linked lists are introduced in the context of an integer stack so that insertion and deletion are only at the front. Next they are used to build a queue that adds support for generics and a tail pointer. Finally, the linked implementation of the list ADT uses double links, a dummy node, and insertion and deletion anywhere in the list. By this point, students have built confidence in their ability to navigate linked list code.

- The organization of topics is designed throughout to reinforce this developmental approach. For example, Chapter 3 develops an **int**-based stack interface, array implementation, and linked implementation. This allows students to concentrate on stack essentials before tackling fully generic stacks in Chapter 4. The List ADT follows stacks and queues for the same reason: students are better prepared to handle its complexity (particularly iterators) after learning stacks and queues first.

- Each section introduces only one or (rarely) two new concepts, allowing students to focus on the new ideas without becoming overwhelmed. This is particularly important during the first half of the course for newcomers to Java.

Other features include:

- Clear, concise explanations of the essential content students need to learn

- An introductory chapter to the basics of Java, allowing students for whom Java is not their first language to quickly get up to speed

- Partial implementations of data structures so that instructors can develop some methods as examples, and students can write other methods as exercises

- Introduction to topics students will likely see later in other courses, such as call stacks, analyzing recursive functions, sorting, inheritance, and abstract classes

- Sections designed to fit approximately one class period each

For the most part, standard Java coding practices are followed, but there are a handful of exceptions:

- Arrays are copied by hand, largely to support learning to resize circular arrays in the queue implementation.

- Java collections are not used, because students are learning to build their own. Most interfaces are close to their `java.util` counterparts, though.

- The unnamed package is used since these applications are all relatively small, and it simplifies file management for beginners.

- The `@Override` annotation is not used for implementations of interface methods, because that use of the term "override" seems likely to confuse beginners.

With the exception of Chapters 10 and 11, which can be interchanged, chapters are designed to be read in order. Section 5.3 should probably not be skipped because it introduces inheritance and the `Object` class.

Java source files and other resources are posted at

`http://www.central.edu/go/datastructures/`

Feedback

Please send any comments or suggestions to johnsonm@central.edu. I look forward to hearing from you.

Acknowledgments

Thanks to all of the students from COSC 130 for making this course so much fun. I hope you can see and appreciate all of the improvements you inspired. Thanks to Randi Cohen, editor at Chapman & Hall/CRC Press, for her ongoing support; I truly appreciate it. And thanks to my colleagues at Central College, including Mark Babcock, director of the college-community chorus, for his regular use of the technique "only one new thing at a time."

About the Author

Mark J. Johnson is professor of computer science and mathematics at Central College in Pella, Iowa, where he holds the Ruth and Marvin Denekas Endowed Chair in Science and Humanities. Mark is a graduate of the University of Wisconsin-Madison (Ph.D., mathematics) and St. Olaf College. He is the author of *A Concise Introduction to Programming in Python*, also published by Chapman & Hall/CRC Press.

Chapter 1

A Brief Introduction to Java

Before beginning our study of data structures, we take a quick tour of the main features of the Java programming language. More advanced features of Java will then be introduced as they are needed. This chapter assumes that you have some programming experience but not necessarily in Java.

A current Java Development Kit (JDK) is required to develop software in Java. All of the examples in this text were written using Java SE 7.

1.1 Basics

We begin with an example that computes factorials in Listing 1.1.

Listing 1.1: Factorial

```java
public class NumericFunctions {
    public static int factorial(int n) {
        int result = 1;
        for (int i = 2; i <= n; i++) {
            result *= i;
        }
        return result;
    }

    public static void main(String[] args) {
        for (int n = 1; n <= 10; n++) {
            System.out.print(n);
            System.out.print(" ");
            System.out.println(factorial(n));
        }
    }
}
```

Although this is a short program, it illustrates a number of features of the Java language. This section is therefore a bit longer than most, so don't necessarily try to remember everything on first reading. Instead, notice what is here, pay attention to things that are different from what you expected, and plan to refer back to this section as questions arise.

The Java Tutorial [13] is a good resource for learning Java, as is *The Java Programming Language* [1]. The authoritative Java reference is the *Java Language Specification* [8].

Classes

Almost all Java code is contained inside class definitions, as in Listing 1.1. Class names are capitalized, and every class is usually stored in its own file, named `ClassName.java`. For now, concentrate on the code inside the two method definitions, `factorial()` and `main()`. Later, in Section 1.5, we will look more closely at the structure of Java class definitions.

Types and Variables

Every variable in Java may only store one **type** of data, and that type must be declared before the variable can be used. A **variable declaration** looks like this:

```
type variableName;
```

Declaration is usually combined with **initialization** to give the variable a value, as in line 3 of Listing 1.1:

```
type variableName = initialValue;
```

This is because **local variables**, which are defined inside blocks such as method bodies, are not given a value until you assign one.

Every declaration limits the scope of the variable that is declared. The **scope** of a variable is the area of a program in which the variable may be used. A variable's scope is limited to the set of braces that its declaration occurs in (known as a block; see below). We generally follow the principle of making the scope of a variable as small as possible; for example, the scope of the variable i in Listing 1.1 is the for-loop in lines 4–6.

Variables that are declared with the **final** modifier are **constant** and can never change once they have been set. Constants are usually written in all capital letters with underscores between words.

Primitive and Reference Types

There are two main kinds of types: primitive and reference. Variables declared with a **primitive** type store their data directly at the variable's location in memory. All other types in Java are **reference** types. References can be thought of as pointers to objects rather than storage for the objects themselves. We concentrate on primitive types in this section, and then gradually introduce reference types throughout the rest of this chapter.

Primitive Types

Java has eight primitive data types, listed in Table 1.1.

TABLE 1.1: Primitive Types

byte	8-bit integer
short	16-bit integer
char	character (16-bit unsigned integer)
int	32-bit integer
long	64-bit integer
float	32-bit floating point
double	64-bit floating point
boolean	**true** or **false**

The most commonly used primitive types are **int**, **double**, and **boolean**. Table 1.2 shows the **literal** values that can be assigned to variables that have been declared with primitive types.

TABLE 1.2: Primitive Literal Values

		Examples
integer	Optional 0x (hex), 0 (octal), or 0b (binary)	173, 0xa5
long	Suffix L	25689L
floating pt	Optional e for base 10 exponent	3.71e8
float	Suffix f	17.0f
boolean	**true** or **false**	
char	Single quotes	'c'

Notice that a stray 0 prefix on an integer constant will cause the value to be interpreted in octal, which is base 8.

Casting can be used to translate between primitive numeric types (excluding boolean). A cast tells the compiler to treat the expression as though it has the specified type. The syntax of a cast is:

```
(type) expression
```

Casting a floating-point value to an integer truncates the fractional part. For example,

```
int i = (int) 7.2913;      // sets i to 7
```

Operators

Java has the standard arithmetic operators +, -, *, and / for numeric types. There is no exponentiation operator. Integer division truncates toward zero, so 5 / 2 is 2 and -5 / 2 is -2. The **remainder** operator is %, defined so that m % n is the remainder after dividing m by n. For example, 5 % 2 is 1. For integer m and n, the following relationship is guaranteed:

```
(m / n) * n + m % n == m
```

Be aware that m % n can be negative: for example, -5 % 2 is -1 because

```
(-5 / 2) * 2 + -5 % 2 == -2 * 2 + -1 == -5
```

The remainder operator is also sometimes called a **modulus** or **mod** operator, but that is technically inaccurate because modulus values are never negative.

Testing equality is done with == or != (not equals). Other **comparisons** are <, >, <=, and >=. Each of these produces a boolean result, either **true** or **false**. Boolean values may be combined with the **boolean operators** && (AND) or || (OR). Any boolean value may be negated by putting a ! (NOT) in front of it.

The boolean operators && (AND) and || (OR) are **conditional**, meaning that they only evaluate the second expression if necessary. For example, if j is negative in this expression:

```
j >= 0 && key < data[j]
```

then the second part, key < data[j] will not be evaluated because the final result is already known to be false. Conditional evaluation is also known as **short-circuiting** the boolean operator.

Increment and Decrement

Numeric variables may be increased by one with the **increment** operator:

```
variable++
```

There is also a **decrement** operator --. Increment and decrement may be used in a stand-alone statement or as part of a larger expression. When present in an expression in the **postfix** form shown above, with the ++ after the variable, the increment or decrement happens *after* retrieving the value of the variable.

In **prefix** form:

```
++variable
```

the increment (or decrement) is done *before* determining the value of the variable.

For example,

```
int x = 10;
int y = x++;
```

changes x to 11 and assigns the value 10 to y because the increment is done after x is evaluated in the second line.

Statements and Blocks

Executable Java code is a sequence of **statements**, each ending with a semi-colon. Braces ({}) may be used to group statements into **blocks**. Braces are also known as squigglies or curly brackets. No semicolons are put after braces.

Blocks usually define the bodies of classes, methods, loops, and if-statements. In general, a block may be used anywhere a single statement can occur—and vice versa—because technically a block is considered a single compound statement. For example, an if-statement may be written with a block:

```
if (test) {
    body;
}
```

or with a single statement:

```
if (test)
    singleStatement;
```

However, because of the danger of adding new statements to the single-statement form without remembering to include the braces, it is better to write single-statement bodies on the same line:

```
if (test) singleStatement;
```

The example statements that follow use the block form.

Assignment

An assignment statement evaluates the expression on the right and assigns that value to the variable on the left:

```
variable = expression;
```

The type of the expression must be compatible with the type of the variable.

Shorthands are available for many common assignments. For example, the `*=` in line 5 of Listing 1.1:

```
result *= i;
```

is an abbreviation of the assignment:

```
result = result * i;
```

Similar shorthands are available for `+=`, `-=`, and `/=`.

Control Statements

Statements are executed in a Java program one after the other, in the order they appear. The **control statements** that follow change the order in which other statements are executed; in other words, they control the flow of execution.

If Statements

An if-statement allows conditional execution by only executing its body if the test is true:

```
if (test) {
    body;
}
```

The `test` inside parentheses must produce a boolean value. There is an optional **else**.

While Loops

A while-loop repeatedly executes its body until the boolean `test` is false:

```
while (test) {
    body;
}
```

For Loops

A for-loop adds initialization and increment steps:

```
for (initialize; test; increment) {
    body;
}
```

This loop works approximately like the following while-loop:

```
initialize;
while (test) {
    body;
    increment;
}
```

The `initialize` step may declare new variables, and any variables declared in this way have their scope limited to the for-loop. The loop at line 4 of Listing 1.1 declares the variable i in this way and executes its body once for each $i = 2, 3, \ldots, n$.

Notice that in a for-loop, the test is checked before the first execution of the body, and the increment step is performed after every execution of the body.

Switch

A switch is an alternative to a long sequence of **if** and **else if** statements.

```
switch (expression) {
    case value1:
        statements
    case value2:
        statements
    ...
    default:
        statements
}
```

The **default** case is optional. When a **switch** executes, control jumps to the matching case if there is one, and then execution continues from that point on, with **case** statements themselves having no effect. In other words, execution does not automatically stop at the end of a case. Thus, most **switch** statements have a **break** statement at the end of every case, in order to cause execution to skip the remaining cases.

Output Statements

Output is generated by the `System.out` functions in Table 1.3.

TABLE 1.3: System.out Print Methods

`System.out.print(item)`
Prints item without appending newline.
`System.out.println(item)`
Prints item, appending newline (\n).

Other output functions offering more flexibility are available; consult the Java documentation or tutorials for details on their use.

Methods

Functions in Java are usually referred to as **methods** because they must be written inside a class. The general format of a method declaration is:

```
modifiers returnType name(parameterType parameter, ...) {
    body;
}
```

The `factorial()` method in Listing 1.1 has modifiers **public static**, return type **int**, and one parameter named n of type **int**. Method modifiers will be discussed in Section 1.5. A method declares its **return type** to indicate the type of data it will return. The **void** return type indicates that no value will be returned. Thus, the `factorial()` method returns an **int**, whereas the `main()` method returns nothing.

Values are returned from non-void methods with a **return** statement:

```
return expression;
```

The type of this expression must be compatible with the method's return type. When a **return** statement is reached, the method immediately stops execution and returns the value of the expression to its caller. Because of this behavior, a **return** statement with no expression may be used in a **void** method to return to the caller without executing any more code within the method.

Method **parameters** such as n in `factorial()`, must declare their type. Parameters are **passed by value**, meaning that the method receives a copy of the value that is passed as the **argument**. For primitive types, this means that a method can modify its copy of the parameter without affecting the caller. Reference parameters will be considered briefly in Section 1.4.

Since Java code needs to be inside a class definition, there is no way to define a function outside of a class, as is common in, say, C or Python. If you find yourself wanting to write a stand-alone function in Java, then generally what you want to write is a static method, such as `factorial()` and `main()` in Listing 1.1. The reason for declaring these methods static will be explained in Section 1.5.

Comments

Java comments come in three varieties:

Multiple line Text contained between /* and */ is ignored by the compiler and meant for others who read your code.

Documentation Multiple line comments beginning with /** are documentation comments that describe the following class or method.

Single line Any text that follows // to the end of the line is treated as a comment, usually to explain a short section of code.

The javadoc tool converts documentation comments into Web pages that follow the same style as the Java library documentation. Documentation comments are not included in this text so that the code itself will be easier to read.

Compiling and Running Java Programs

To run Listing 1.1, save the code in a file named NumericFunctions.java. This Java **source code** is compiled to Java bytecode with the command:

```
% javac NumericFunctions.java
```

Bytecode is like a low-level machine language, except that it consists of instructions for the **Java Virtual Machine** or **JVM** rather than a particular CPU. The compiler stores the generated bytecode in a file named NumericFunctions.class. Once this bytecode file has been created, the program may be executed with the command:

```
% java NumericFunctions
```

This will automatically call the main() method from the NumericFunctions class, as long as that method is declared **public static void** with a String[] parameter, as on line 10. **Integrated development environments** (IDEs) such as Netbeans or Eclipse make it easy to edit, compile, and run Java programs without having to work from the command line. jGRASP [7] is an IDE specifically designed for learning data structures.

Exercises

1. Suppose the **int** variables x and y have the values 5 and 10, respectively. Give the value of both x and y after each of these Java statements runs, starting over with the original values of x and y each time:

   (a) x = ++y;   (c) x = y++;

   (b) x = --y;   (d) x = y--;

2. Write for-loops to produce the following values:

    (a) $i = 1, 2, 3, \ldots, 100$

    (b) $j = 0, 1, 2, 3, \ldots, 9$

    (c) $m = 0, 2, 4, 6, \ldots, 100$. (Hint: the increment step can be a shorthand assignment.)

    (d) $n = 1, 2, 4, 8, \ldots, 1024$.

3. Write three different for-loops that will each execute their body ten times. Do not just change the name of the loop variable; change all three parts of the for-loop.

4. Modify `main()` in Listing 1.1 to include a call to `factorial(0)`. Explain why the correct result $(0! = 1)$ is returned.

5. Modify `main()` in Listing 1.1 to include calls to `factorial(n)` with negative n. Does the program crash? Explain its behavior, focusing on the for-loop on line 4.

6. Modify `main()` in Listing 1.1 to compute factorials up to $n = 20$.

    (a) Determine the largest n for which the correct value is returned.

    (b) Explain why some incorrect values are computed. Hint: a 32-bit signed **int** can hold values between -2^{31} and $2^{31} - 1$.

    (c) Modify the `factorial()` function to use the **long** data type so that larger factorials can be accurately computed. Determine the largest n for which the correct value is returned by your modified function.

7. Write a Java function `pow(m, n)` that computes m^n for integers m and n, assuming $n \geq 0$. Use a loop to repeatedly multiply by m. Add the `pow()` function to the `NumericFunctions` class and include code in `main()` to compute `pow(m, n)` for all m and n between 1 and 9.

8. Write a Java function to implement **Euclid's algorithm** for computing the greatest common divisor `gcd(m, n)`, which is the largest integer k dividing both m and n. One form of Euclid's algorithm is:

```
while n > 0
    replace n with m % n
    replace m with the previous value of n
```

When the loop stops, the gcd is in m. Add the `gcd()` function to the `NumericFunctions` class and include code in `main()` to compute `gcd(m, n)` for all m and n between 2 and 10.

1.2 Strings

In this section, we look at an example in Listing 1.2 featuring the Java `String` type. Although we will only use strings occasionally, they provide a natural first example of a reference type in Java.

Listing 1.2: Count Substrings

```java
public class StringFunctions {
    public static int count(String s, String target) {
        int count = 0;
        int n = target.length();
        for (int i = 0; i <= s.length() - n; i++) {
            String piece = s.substring(i, i + n);
            if (piece.equals(target)) count++;
        }
        return count;
    }

    public static void main(String[] args) {
        System.out.println("Number of this's: " +
            count("this and this and that and this", "this"));
    }
}
```

The `String` class in Java provides support for working with character text. As you may have noticed, **string literals** are contained in double quotes.

Concatenation

Strings may be combined via **concatenation** using + or the shorthand +=. Thus, the sequence

```java
String s = "abc";
s += "def";
```

results in s having the value `"abcdef"`.

However, strings are **immutable**, meaning that once a `String` has been created its contents can never change. What happens behind the scenes, then, when executing

```java
s += "def"
```

is that a new `String` object is created with the value `"abcdef"`, and the reference s is changed to point to this new string instead of the old one that contained "abc". Thus, string concatenation is relatively inefficient because it repeatedly generates new `String` objects. We will see a more efficient technique for building strings in Section 1.4.

Concatenation may also be used to combine strings with other types that are capable of being converted to strings. This conversion is done automatically, as in lines 13 and 14 of Listing 1.2. In that example, the **int** returned by `count()` is automatically converted to a string so that it can be concatenated with the literal string `"Number of this's: "`.

String Methods

Recall that primitive types provide storage for a simple piece of data, such as an integer or float. References that point to objects are capable of more advanced behaviors, including method calls. We will discuss general object behavior in Section 1.4; for now, Listing 1.2 demonstrates calling a few of the string methods from Table 1.4.

TABLE 1.4:   String Methods

char `charAt(int i)` Character at position i.
int `indexOf(String s)` Index of first occurrence of s in this string, -1 if not found.
int `indexOf(String s, int start)` Index of first occurrence of s in this string starting at index start, -1 if not found.
int `length()` Number of characters in string.
`String substring(int i)` Substring starting at index i.
`String substring(int i, int j)` Substring from index i to j - 1.
`String toLowerCase()` Returns copy in all lowercase.
`String toUpperCase()` Returns copy in all uppercase.
`String trim()` Returns copy with whitespace removed from each end.

Strings can be thought of as sequences of characters, indexed beginning at 0, such as

$$\begin{array}{cccccccc} \underline{0} & \underline{1} & \underline{2} & \underline{3} & \underline{4} & \underline{5} & \underline{6} & \underline{7} \\ A & & s & t & r & i & n & g \end{array}$$

Several of the methods in Table 1.4 are based on this indexing.

It is important to notice that methods whose names sound like they might change the string (`toLowerCase()`, `trim()`, etc.) *do not* change the string they are called on. This is because strings are immutable; they never change. Instead, such methods always return a modified copy of the original string.

Consult the Java API documentation [10] for a complete list of string methods.

Comparing Strings

In general, reference types must be compared differently than primitives, both for equality and inequalities. We will learn about those differences gradually, but for now, we concentrate on how to compare strings. Table 1.5 lists the main string comparison methods.

TABLE 1.5: String Comparisons

`int compareTo(String s)`   Returns negative if this string comes alphabetically before s, zero if equal, and positive if alphabetically after.
`int compareToIgnoreCase(String s)`   Same as `compareTo()` except ignores upper and lowercase.
`boolean equals(Object o)`   Returns true if o is a string with the same contents as this string.
`boolean equalsIgnoreCase(String s)`   Returns true if s has the same contents as this string, ignoring case.

There are two basic rules for comparing strings:

Use equals() if you want to know if two strings have the same contents. Do not use ==. Notice that line 7 in Listing 1.2 uses `equals()` to compare each piece with the target.

Use compareTo() if you want to compare strings with respect to their alphabetical order. Do not try to use <, >, or other numeric comparisons.

Be careful: the compiler will allow you to use == with strings, but that will almost never be what you want. On the other hand, the compiler will not allow you to use other comparisons with strings.

The `compareTo()` method is based on the ASCII codes of the characters in the strings, and so uppercase letters precede their lowercase equivalents. For example,

```
"apple".equals("banana")                  \\ false
"apple".compareTo("banana")               \\ negative
"apple".compareTo("BANANA")               \\ positive
"apple".compareToIgnoreCase("BANANA")     \\ negative
```

Do not worry about the precise values returned by the `compareTo()` methods: in programs, you will just need to test if values are positive or negative.

The difference between `equals()` and `==` applies to all Java objects, not just strings, and so we will come back to that distinction in Section 1.4. Later, in Section 5.3, we will explain the reason that the parameter to `equals()` is an `Object`. For now, just think of it as a second string.

Exercises

1. Suppose s = `"data structures"`. Give the value of each of these expressions:

   (a) `s.length()`   (g) `s.indexOf("t", 7)`

   (b) `s.charAt(5)`   (h) `s.indexOf("d", 0)`

   (c) `s.indexOf("a")`   (i) `s.indexOf("d", 1)`

   (d) `s.indexOf("m")`   (j) `s.substring(5)`

   (e) `s.indexOf("struct")`   (k) `s.substring(1, 3)`

   (f) `s.indexOf("t", 3)`   (l) `s.toUpperCase()`

2. Suppose t = `"Java programming language"`. Give the value of each of these expressions:

   (a) `t.length()`   (g) `t.indexOf("g", 16)`

   (b) `t.charAt(0)`   (h) `t.indexOf("r", 6)`

   (c) `t.indexOf("s")`   (i) `t.indexOf("r", 7)`

   (d) `t.indexOf("n")`   (j) `t.substring(5)`

   (e) `t.indexOf("gram")`   (k) `t.substring(5, 12)`

   (f) `t.indexOf("g", 9)`   (l) `t.toLowerCase()`

3. Suppose s = `"stack"`, t = `"queue"`, and u = `"Stack"`. Give the value of each of these expressions:

   (a) `s.equals(t)`   (c) `!t.equalsIgnoreCase(u)`

   (b) `!s.equals(u)`   (d) `s.equalsIgnoreCase(u)`

4. Suppose s = `"stack"`, t = `"queue"`, and u = `"Stack"`. Decide whether each of these is positive, negative, or zero:

   (a) `s.compareTo(t)`

   (b) `s.compareTo(u)`

   (c) `t.compareToIgnoreCase(u)`

   (d) `s.compareToIgnoreCase(u)`

5. Suppose s is a reference to a Java string. Write expressions to return each of these values:

   (a) The length of s

   (b) The third character in s (as in, the third character of "abcd" is "c")

   (c) The substring of s consisting of its third through fifth characters

   (d) The substring of s consisting of its fourth character to the end

6. Suppose s and t are references to Java strings.

   (a) Write an if-statement that will execute its body only when s and t have the same contents.

   (b) Write an if-statement that will execute its body only when s and t have different contents.

   (c) Write an if-statement that will execute its body only when s comes alphabetically before t.

7. Modify line 7 of Listing 1.2 to use == instead of `equals()`. Describe the results.

8. Modify Listing 1.2 to write a method called `countIgnoreCase()` that ignores case. Test your method in `main()`.

9. Write a version of the `count()` method from Listing 1.2 called `count2()` that uses a while-loop and the `indexOf()` string method to search for the target. Use the version of `indexOf()` with two parameters. Add your function to the `StringFunctions` class and test your method in `main()`.

10. Determine how the value of the string `compareTo()` method is computed, either through experimentation or by consulting the Java API documentation.

1.3   Arrays

Arrays are a simple data structure that provide a basis for creating many of the other, more complex structures we will study. They also provide a second example of a reference type. Listing 1.3 demonstrates the use of arrays with a standard search technique known as linear search.

An **array** is an ordered collection of variables, all of the same type. A reference to an array then allows accessing any variable in the collection by its index.

Listing 1.3: Linear Search

```java
public class ArrayFunctions {
    public static int linearSearch(int[] data, int target) {
        for (int i = 0; i < data.length; i++) {
            if (target == data[i]) return i;
        }
        return -1;
    }

    public static void main(String[] args) {
        int[] data = {3, 14, 7, 22, 45, 12, 19, 42, 6};
        System.out.println("Search for 7: " +
                linearSearch(data, 7));
        System.out.println("Search for 100: " +
                linearSearch(data, 100));
    }
}
```

Declaring Array References

The declaration:

```
type[] arrayRef;
```

declares `arrayRef` to be a reference to an array in which each variable has the specified type. Thus, `data` is declared to be an array of **int** variables in line 10 of Listing 1.3.

No actual array is created by declaring a variable with an array type. This is because array types are reference types, and so declaring an array type variable simply sets aside storage for this variable to eventually hold a reference to an actual array. Creating the array is a separate step.

Creating Arrays

There are two main ways to create an array: with an initializer or using **new**. An **initializer** is appropriate when the array contains a small number of known items, as in line 10 of Listing 1.3. The syntax of an array initializer is:

```
arrayRef = {item, item, ..., item};
```

More often, **new** is used with a type and square brackets to create an array of a given size:

```
arrayRef = new type[N];
```

Here the `type` used with **new** must be compatible with the type that was used to declare the array variable, and N specifies the size of the new array, also known as its **length**. An array created with **new** begins with the default initial value at each index. For numeric types, the default is zero; for reference types the default is the **null** reference (see Section 1.4).

The length of an array is accessible in its `length` field:

```
arrayRef.length
```

See, for example, line 3 in Listing 1.3. The length of an array is fixed at the time it is created and can never change. However, an array variable can be reassigned at any time to point to a different array object that has a different length.

Accessing Array Elements

Items in an array are accessed by their **index** or location in the array. Indices are numbered starting at 0, so an array with 9 items in it such as `data` in Listing 1.3 has indices 0 through 8. We visualize the `data` array like this:

0	1	2	3	4	5	6	7	8
3	14	7	22	45	12	19	42	6

The item at index i in an array is accessed using square brackets:

```
arrayRef[i]
```

For example, `data[2]` is 7, and `data[7]` is 42.

Java performs automatic **bounds checking** on all array accesses. If an invalid index is used, an `ArrayIndexOutOfBoundsException` is thrown, generally stopping execution of your program.

Enhanced For-Loop

Java collections, including arrays, support a simplified **for** statement that allows looping over each item in the collection. Known as the **enhanced for-loop**, the syntax is:

```
for (type variable : collection) {
    body;
}
```

This loop will execute once for each item in the collection. Inside the body of the loop, each item is stored in `variable`. The `type` specified for the variable has to match the type of each item in the collection.

For example, consider the `data` array from the `main()` method in Listing 1.3. It is declared as type `int[]`, and so the item type for an enhanced for-loop using that array should be `int`. This loop prints all of the items in the array:

```
for (int item : data) {
    System.out.println(item);
}
```

The enhanced for-loop is quite nice in situations when you do not otherwise need the array indices.

Linear Search

The method in Listing 1.3 solves the **search problem**:

> Given an array `data` and a particular value `target`, determine whether or not `target` appears as an element in `data`. If it does, return the smallest index where it occurs; otherwise, return -1.

We solve the search problem by producing an algorithm. An **algorithm** is a specific sequence of computable steps designed to solve a particular problem. Notice that this definition contains several vague terms; for our purposes, an algorithm will simply be a program that solves a problem.

Linear search, also known as **sequential search** solves the search problem by examining each element in the array, one at a time beginning at item 0, keeping track of the current index. If the item is found, the corresponding index is returned; otherwise, if after looking at every element the item has not been found, it returns -1.

Because the problem statement specifies returning the index where the element is found, we need to loop over indices rather than, say, the elements in the array.

Exercises

1. Write a Java statement to declare each of these array references:

   (a) An **int** array named `counts`
   (b) A **double** array named `times`
   (c) A **boolean** array named `visible`
   (d) A String array named `names`

2. Use an initializer to write a Java statement to declare and create each of these arrays:

   (a) An **int** array named `counts` with contents 18, 3, 9, 22, 11, 4.
   (b) A **double** array named `times` with contents 1.52, 1.98, 1.44, 1.63, 1.67.
   (c) A **boolean** array named `visible` with contents T, F, F, T, F.
   (d) A String array named `names` with contents "Alice," "Bob," "Carol," "Dave."

3. Write a Java statement to declare each of these array references and create an array of the given length that the variable refers to.

   (a) An **int** array named `counts` of length 10
   (b) A **double** array named `times` of length 40
   (c) A **boolean** array named `visible` of length 1000
   (d) A String array named `names` of length 100

4. Write an enhanced for-loop to print every item in each of these arrays:

   (a) An **int** array named `counts`
   (b) A **double** array named `times`
   (c) A **boolean** array named `visible`
   (d) A String array named `names`

5. Use Listing 1.3 to:

   (a) Determine the name and type of the one parameter to `main()`.
   (b) Explain why the test in the for-loop on line 3 uses less than rather than less than or equal to.
   (c) Trace the execution of the first call to `linearSearch()` on line 12.
   (d) Trace the execution of the second call to `linearSearch()` on line 14.

6. Explain why an enhanced for-loop is not used in the `linearSearch()` method of Listing 1.3.

7. Explain why it works to use == in line 4 of the `linearSearch()` method. Would substituting `equals()` also work?

8. Add an enhanced for-loop to `main()` in Listing 1.3 that searches for every element in the array `data`.

9. Write a `sum(int[] data)` method for the `ArrayFunctions` class that returns the sum of the elements in the given array. Test your implementation in `main()`.

10. Write a `max(int[] data)` method for the `ArrayFunctions` class that returns the value of the largest element in the given array. Assume the array is nonempty, and test your implementation in `main()`.

11. Write a `min(int[] data)` method for the `ArrayFunctions` class that returns the value of the smallest element in the given array. You may assume the array is nonempty. Test your implementation in `main()`.

12. Write a `display(int[] data)` method for the `ArrayFunctions` class that prints the contents of the given array on one line. Use `print()` for each element, but finish with a final `println()` so that future output appears on its own line. Test your implementation in `main()`.

1.4   Using Objects

Java is an **object-oriented** programming language, and so, learning to use objects is one of the keys to successfully adapting to the language. You have seen two types of objects so far, strings and arrays. Each of those is somewhat unique; this section uses a more representative example to convey the general ideas.

Objects are defined by classes. In this section, we focus on how to use objects; in the next section, we look at how to write our own classes. In addition to using a new object, Listing 1.4 ties together some ideas from the last two sections. Before looking at the details of how the `StringBuilder` class and the `split()` method are used in it, we consider some general issues involved in using Java objects.

Declaring Object References

Object references are declared in the same way as primitive variables:

```
type referenceVariable;
```

Listing 1.4: Acronym

```
1  // Add to class StringFunctions:
2
3  public static String acronym(String phrase) {
4     StringBuilder result = new StringBuilder();
5     for (String token : phrase.split("\\s+")) {
6        result.append(token.toUpperCase().charAt(0));
7     }
8     return result.toString();
9  }
```

The type that a reference is declared with defines the methods and fields that may be used via the reference. However, as with strings and arrays, declaring a reference variable does not create an object for the variable to refer to, and a reference cannot "do" anything until it refers to an object. Once an object has been created, we often talk about the reference variable as if it is the object that it references.

Creating Objects

Objects are created in Java using **new** and the class constructor:

```
referenceVariable = new ClassName(arguments);
```

Constructors always have the same name as their class, and may take zero or more parameters. In Listing 1.4, a `StringBuilder` object is created by calling the constructor on line 4, and that object is then referenced by the variable `result`.

Accessing Fields and Calling Methods

Objects store their data in **fields** defined by the class from which they were created. Fields are accessed using **dot notation**:

```
object.field
```

We saw this notation in the last section for accessing the `length` field of an array.

Methods are operations that may be performed on or with the data stored in the object's fields. Methods are **called** using the same dot notation:

```
object.method(arguments)
```

There were several examples of `String` method calls in Section 1.2, and line 6 of Listing 1.4 calls the `append()` method on the `StringBuilder` object `result`. It is also worth noting that `print()` and `println()` are method calls on the `System.out` object.

Accessing a field or method of a reference variable using a dot is sometimes called **dereferencing** the variable. Only variables declared with reference types can be dereferenced. If you try to dereference a variable declared with a primitive type, you well get an error.

Object References

Recall that reference types point to the object they refer to using a **reference**, rather than directly storing data like primitives. Given these assignments:

```
int x = 17;
String s = "This is a string";
```

it is helpful to imagine the difference like this:

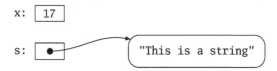

because the actual contents of `s` are a reference to the string object. Java references are also called **pointers** because they point to objects; be aware, however, that in other languages there may be differences between pointers and references. If another reference is created and given the same value as `s`:

```
String t = s;
```

then both references point to the same object:

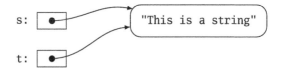

Recall (page 8) that Java passes all method parameters by value. This means that for object references, parameters always get a copy of the reference (in the same way that `t` is a copy of the reference `s` in our example here), but the reference points to the actual object. Generally, this is exactly what you expect and need.

There is one literal reference, **null**, which refers to no object. Trying to access a field or method of a null reference will generate a `NullPointerException`.

Using equals()

We can say a bit more now about the difference between == and `equals()` for reference types. With all object types (including strings), == and != check whether or not the two references refer to the same object. The `equals()` method, on the other hand, is defined by classes such as `String` to check whether or not the objects have the same content.

Thus, most of the time, when comparing objects, you want to compare their contents and therefore use `equals()`. Direct comparisons with == are used mostly with primitive types.

There is one caveat: Java arrays *do not* implement `equals()` to compare their contents. If you need to compare the contents of arrays, use the `equals()` method from the `java.util.Arrays` class.

Destroying Objects

There is no explicit way to destroy an object in Java after it has been created. Instead, Java has a system of automatic **garbage collection** that is allowed to reclaim memory once it is no longer being used. The garbage collector determines that an object is no longer being used when there is no way to reach it by following active references in the program. This automatic system is nice; other languages may require you to take out your own trash.

However, there still may be times when we need to eliminate references to objects that are no longer being used. In general, if x is a reference to an object we no longer need, then setting x to **null** will alert the garbage collector that the object can be deleted (assuming no other references point to it).

StringBuilders

Recall that string concatenation is inefficient because strings are immutable (see Section 1.2). `StringBuilder` objects are specifically designed to perform efficient string accumulation. Because they are **mutable**, their contents can change and so they do not require creating new objects for every concatenation.

The `StringBuilder` object referenced by `result` in Listing 1.4 is created on line 4 by calling a constructor with **new**. Once the object exists, the `StringBuilder` methods listed in Table 1.6 can be called on it.

TABLE 1.6: StringBuilder Methods

void append(String s)
Appends s to current contents.
String toString()
Converts contents to string.

String Tokenizing

A common task is to break a given string into a collection of substrings called **tokens**. In Listing 1.4, we need each word in the phrase in order to be able to create the acronym. In that case, the tokens are the individual words, separated by whitespace.

The `String split()` method, given in Table 1.7, provides a convenient method for tokenizing strings, returning a `String` array that contains the tokens. The `split()` method breaks a string into tokens at each occurrence of the expression passed as its parameter. For example,

```
"Name:Address:City:State:Zip".split(":")
```

returns the array

```
{"Name", "Address", "City", "State", "Zip"}
```

The split expression `":"` is not kept as part of any of the tokens.

TABLE 1.7:   String Tokenizing

`String[] split(String expr)`
Returns an array of strings created by splitting this string at each occurrence of `expr`.

The split expression can be any **regular expression**. The regular expression used in Listing 1.4 matches any string of one or more characters of **whitespace** (spaces, tabs, or newlines). It works like this: the pattern "\s" matches any single whitespace character. But to use the pattern "\s" in a string literal (as here), the backslash must be **escaped** with another backslash; thus the string literal pattern `"\\s"` matches any one character of whitespace. Appending a "+" to a pattern matches one or more occurrences of the pattern, and so, `"\\s+"` matches any set of one or more characters of whitespace.

Exercises

1. Suppose `Random` is the name of a Java class with a constructor that takes no parameters. Write a Java statement to declare and create a `Random` object named `gen`.

2. Suppose `Thread` is the name of a Java class with a constructor that takes no parameters. Write a Java statement to declare and create a `Thread` object named `t`.

3. Suppose `Button` is the name of a Java class with a constructor that takes a string parameter for the button's label. Write a Java statement to declare and create a `Button` object named `goButton` labeled "Go."

4. Determine the values of these Java expressions:

    (a) `"This    sentence has  long  gaps.".split("\\s+")`

    (b) `"What about punctuation?".split("\\s+")`

    (c) `"This is a sentence.".split("\\s")`

    (d) `"This is a sentence.".split("s")`

    (e) `"This is a sentence.".split("e")`

5. Add the `acronym()` method from Listing 1.4 to the `StringFunctions` class. Include code in `main()` to test it.

6. Rewrite the `acronym()` method from Listing 1.4 to use a regular indexing for-loop instead of an enhanced for-loop. Name the method `acronym2()` and test it in `main()`.

7. Write a method `countWord(String s, String target)` that counts the number of times the word `target` occurs in the string `s`. (Assume `s` contains no punctuation so that words are delimited by whitespace.) Add your method to the `StringFunctions` class and test it in `main()`.

8. Write a method `countWordIgnoreCase(String s, String target)` that counts the number of times the word `target` occurs in the string `s`, ignoring case. (Assume `s` contains no punctuation so that words are delimited by whitespace.) Add your method to the `StringFunctions` class and test it in `main()`.

9. Write a Java method `reverse(String s)` for the `StringFunctions` class that returns a string consisting of the characters in `s` in reverse order. For example, `reverse("abc")` should return the string `"cba"`. Test your method in `main()`.

10. Add a new `linearSearch()` method to the `ArrayFunctions` class that searches for a target string in an array of strings. Test your method in `main()`. (Note: Java allows multiple versions of the same method as long as they have different lists of parameter types; we will explain this feature in the next section.)

1.5 Writing Classes

Classes allow us to define new data types in Java. Types defined by classes are known as **object types** or **class types**, and all object types are reference types. Listing 1.5 defines a `Fraction` type.

Listing 1.5: Fraction Class

```java
public class Fraction {
    private int num;
    private int den;

    public Fraction(int num, int den) {
        this.num = num;
        this.den = den;
    }

    public Fraction(int n) {
        this(n, 1);
    }

    public void addOn(Fraction f) {
        num = num * f.den + den * f.num;
        den *= f.den;
    }

    public static Fraction add(Fraction f1, Fraction f2) {
        int n = f1.num * f2.den + f1.den * f2.num;
        int d = f1.den * f2.den;
        return new Fraction(n, d);
    }

    @Override
    public String toString() {
        return num + "/" + den;
    }

    public static void main(String[] args) {
        Fraction f1 = new Fraction(3, 4);
        Fraction f2 = new Fraction(1, 3);
        System.out.print(f1 + " + " + f2 + " = ");
        System.out.println(add(f1, f2));
        f1.addOn(f2);
        System.out.println("Using addOn() changes f1 to " + f1);
    }
}
```

Class Definitions

A typical Java class definition looks like:

```
modifiers class ClassName {
    // field declarations
    // method definitions
}
```

The body of the class inside its braces defines the class **members**, usually fields and methods. It is also possible to define classes within classes, known as nested classes. We will see one use for this later when creating linked structures.

Visibility

Java allows you to specify the visibility of a class and its members with **access modifiers**. These modifiers control what other sections of code are able to use the class or its members. If a class is declared **public**, the class may be used by any other class. In general, a public class should be stored in a file with the same name as the class.

For now, we will use two access modifiers for class members:

> **public** Available anywhere the class is available.
> **private** May only be accessed inside the class itself.

Protected access will be discussed in Section 5.3; package visibility will not be used in this text.

The primary use of private visibility is to provide **encapsulation** for objects, giving them an outer shell that other classes cannot break through. This allows each class to manage its own objects by limiting other classes to using only publicly declared members.

Declaring and Initializing Fields

A **field declaration** is simply a variable declaration inside the body of a class that is not contained in the body of any method definition. For example, lines 2 and 3 of Listing 1.5 define the two fields num and den in the Fraction class. Fields are normally declared private so that other classes are forced to use public methods to access class or object data.

Fields may be initialized at the time of declaration, or they may be assigned values by a constructor. If not initialized, fields are given a **default value** based on their type: generally, numeric types are set to 0 and object references are null. The fields num and den are not initialized, so they are given default values which will be reset by the constructor (see below).

Writing Methods

You are already familiar with writing methods in classes, although prior to this section, all of our methods have been static. As described in Section 1.1, methods must declare their return type and the type of their parameters. Each method may also declare its own visibility with an access modifier.

Inside non-static methods, the keyword **this** refers to the current object on which the method was called. Thus, inside method bodies (including constructors), this syntax:

```
this.fieldName
```

references a field rather than some other variable such as a parameter or local variable. In lines 6 and 7 of Listing 1.5, this allows the parameters of the constructor to have the same names as the fields, instead of requiring the programmer to choose different names for the parameters.

For example, in line 6:

$$\text{\textbf{this}.num} = \text{num};$$
$$\uparrow \qquad \uparrow$$
$$\text{field} \quad \text{parameter}$$

this.num on the left refers to the field named num, whereas num on the right refers to the parameter. This is generally the only time we reuse field names; otherwise, there is a danger that the fields will be hidden by local variables or parameters. There is no need to use **this** if there is no name conflict.

Writing Constructors

Recall from the previous section that constructors are used with **new** in order to create new objects. Thus, the primary purpose of a constructor is to set up new objects in a valid **state**, meaning that its fields have meaningful values. In the case of the Fraction class, there are no natural default values for the numerator and denominator, so values for the numerator and denominator must be specified when the fraction is created. Thus, instead of using field initializers, we write constructors to provide convenient means for fraction objects to be created.

Constructors always have the same name as the class and do not have a return type. They may specify their own visibility.

Overloading Methods and Constructors

Java allows defining more than one version of a method or constructor as long as each version has a unique signature. The **signature** of a method is its name and list of parameter types. Neither the return type nor parameter names are part of a signature. Defining multiple versions of a method is called **overloading** the method.

In Listing 1.5, the constructor is overloaded; the signatures of the two constructors are:

```
Fraction(int, int)
Fraction(int)
```

When overloading constructors, **this()** may be used as the first statement of a constructor in order to call a different constructor:

```
this(arguments);
```

This reduces duplication of code, usually by writing full code for the most general constructor and then calling **this()** in the more specialized constructors. For example, in Listing 1.5, the first constructor is a general constructor for fractions n/d, while the second constructor calls **this(n, 1)** on line 11 to create fraction objects for integers $n = n/1$.

Instance and Static Methods

The syntax for method calls described in Section 1.4 is the standard way of calling **instance methods**, which are always called from an **instance** of a class, meaning an object:

```
object.method(arguments)
```

Instance methods have no special designation in a class; in other words, all methods are instance methods unless declared otherwise. In Listing 1.5, the addOn() and toString() methods are instance methods. You can see the above calling syntax used with the addOn() method in line 35. There, the addOn() method is called on the instance f1, modifying f1 by adding f2 onto it.

In contrast, **static methods** do not need to be called from any object.[1] They are considered **class methods** rather than object methods. In fact, you can think of "static" as a synonym for "class." Methods are declared static with the **static** modifer:

```
static returnType staticMethod(parameters) { ... }
```

Inside of its own class definition, a static method may be called by simply providing the name of the method and its arguments:

```
staticMethod(arguments)
```

---

[1] They may be called from objects, but we will have no need to do that.

This syntax is used in the call to `add()` in line 34, where the sum of `f1` and `f2` is printed.

Compare the `add()` and `addOn()` methods to help understand the difference between instance and static methods:

addOn() is an instance method so must be called from a particular fraction. It modifies the fraction it is called on.

add() is a static method and adds two given fractions, returning a new fraction as the result.

Because instance methods are more common than static methods, the term "method" will normally refer to instance methods.

Calling Static Methods from Other Classes

Static methods from other classes are called by prefixing the method name with the class name:

```
ClassName.staticMethod(arguments)
```

Instance and Static Fields

As with methods, there are also instance fields and static fields. An **instance field** or **instance variable** is a variable for which each object has its own separate storage. There is no special designation for instance fields, so once again, all fields are instance fields unless specified static. Thus, in the `Fraction` class,

```
private int num;
private int den;
```

both `num` and `den` are instance variables, and so every fraction stores its own numerator and denominator.

Fields may be declared **static**, in which case the entire class shares a single storage location for that field. Because of this, static fields can be thought of as **class variables**. Static fields are not used as often as instance fields, so the term "field" usually refers to instance variables.

toString() Methods

Any class may provide a `toString()` method to indicate how objects of that type should be converted to strings. In Listing 1.5, the `toString()` method of the `Fraction` class returns the string

```
num + "/" + den
```

One of the advantages of writing a `toString()` method is that it will be called automatically by `System.out.println()` when printing an object of that type. Thus, for example, when `f1` is printed at line 33, its output will be "3/4."

We will revisit `toString()` methods and the `@Override` annotation used in Listing 1.5 in Section 5.3.

Exercises

1. Give the signatures of each of these methods from Listing 1.5:

   (a) `addOn()`   (c) `toString()`

   (b) `add()`   (d) `main()`

2. Give the signatures of each of these methods:

   (a) `factorial()` from Listing 1.1

   (b) `count()` from Listing 1.2

   (c) `linearSearch()` from Listing 1.3

   (d) `acronym()` from Listing 1.4

3. Explain why the methods listed in Exercise 2 were declared static.

4. Would Java allow the name of the `addOn()` method in Listing 1.5 to be changed to `add()`? Explain why or why not.

5. Add code to the `main()` of Listing 1.5 to accomplish these tasks:

   (a) Declare and create a `Fraction` named f with value 5/8. Print the value of f.

   (b) Declare and create a `Fraction` named g with value 17/3. Print the value of g.

   (c) Declare and create a `Fraction` named h with value f + g. Print the value of h.

   (d) Declare and create a `Fraction` named j with value 5. Print the value of j.

   (e) Increase the value of g by 5. Print the new value of g.

6. Add these methods to the `Fraction` class of Listing 1.5. Modify `main()` to test your functions.

   (a) `subtractOff(Fraction f)` to subtract f from this fraction

   (b) `multiplyBy(Fraction f)` to multiply this fraction by f

   (c) `divideBy(Fraction f)` to divide this fraction by f

   (d) `addOn(int n)` to add the integer n to this fraction

7. Add these methods to the `Fraction` class of Listing 1.5. Modify `main()` to test your functions.

    (a) `subtract(Fraction f1, Fraction f2)` to compute f1 - f2

    (b) `multiply(Fraction f1, Fraction f2)` to compute f1 * f2

    (c) `divide(Fraction f1, Fraction f2)` to compute f1 / f2

8. Write a `reduce()` method for the `Fraction` class of Listing 1.5 that reduces this fraction to its lowest terms. Assume all terms are positive and use the `gcd()` function from Exercise 8 in Section 1.1. Add calls to the `reduce()` function to other methods in the class where they are necessary.

9. Modify the `reduce()` method from the previous exercise to handle negative fractions correctly. Have the constructor make sure all denominators are positive, so that negative fractions have a negative numerator. Use the `Math.abs()` library function if you need absolute value.

Chapter 2

Algorithm Analysis

An important part of learning about data structures is understanding their performance. There are at least two reasons:

Design should take performance into account. What might seem like a natural and easy to code design might be unacceptably slow.

Selection of a data structure for a particular task should also take performance into account.

As you will see, data structures often get fast performance for some operations by sacrificing the performance of others. Being aware of these tradeoffs is part of becoming a good software developer.

There are also two main aspects to algorithm performance: time and space. Time performance is simply how long it takes the algorithm to run, whereas space is concerned with how much memory the algorithm requires. We will concentrate mostly on time.

2.1   Big-O Notation

Big-O notation is an example of an **asymptotic** notation that captures the overall behavior of an algorithm or mathematical function for large input values. In order to get a feel for the usefulness of asymptotic notations, it will help to look at an example.

Suppose we have three algorithms, A, B, and C, which take different numbers of steps depending on their input size n:

Algorithm	Number of Steps
A	$10n + 25$
B	$0.1n^2 + n + 28$
C	n^2

If we compute the number of steps each algorithm takes for different n, the results look like this:

Input Size	Algorithm A	Algorithm B	Algorithm C
10	125	48	100
20	225	88	400
40	425	228	1600
80	825	748	6400
160	1625	2748	25600
320	3225	10588	102400
640	6425	41628	409600
1280	12825	165148	1638400
2560	25625	657948	6553600
5120	51225	2626588	26214400
10240	102425	10496028	104857600

It appears that over the long run, algorithms B and C behave similarly, and that algorithm A is faster than both of them.

We can extract these principles from the table:

Focus on large n Algorithm B is fastest for $n < 100$, but that is not a good indicator of its overall performance. It is much slower than A for large n.

Focus on highest power The columns for n and algorithm A grow very similarly, as do the columns for B and C. What these pairs have in common is their highest power of n: for n and A it is n iteself; for B and C it is n^2.

Ignore other detail Compare the growth of algorithms B and C. The coefficients and smaller terms in $0.1n^2 + n + 28$ do not create an important difference with n^2 itself.

To be realistic, the table should have continued for much larger n, into the millions and tens of millions. You can imagine what that would look like.

Big-O Notation

Big-O notation captures this overall behavior of an algorithm or mathematical function for large n. We say that **g(n) is O(f(n))** if there is a constant c so that for all large n,

$$g(n) \leq c \cdot f(n)$$

An algorithm is considered $O(f(n))$ if it takes $g(n)$ steps and $g(n)$ is $O(f(n))$. However, rather than using this precise definition, we will apply the principles derived above to focus on highest powers and ignore other detail.

For example,

Algorithm A is O(n) because $10n + 25$ is $O(n)$. The highest power is n and we ignore the coefficient 10 and lower term.

Algorithm B is O(n²) because $0.1n^2 + n + 28$ is $O(n^2)$. The highest power is n^2 and we ignore the coefficient and both lower terms.

Algorithm C is O(n²) because n^2 is itself $O(n^2)$.

Notice how $O()$ captures the sense we had of which algorithms had the "same" performance.

Analyzing an algorithm's running time using $O()$ notation is known as studying its **time complexity**, while **space complexity** describes memory requirements. Most algorithms have one of the time complexities in Table 2.1, ordered from fastest to slowest.

TABLE 2.1: Common Time Complexities

$O(1)$	Constant
$O(\log n)$	Log
$O(n)$	Linear
$O(n \log n)$	
$O(n^2)$	Quadratic
$O(n^3)$	Cubic

What to Count

Depending on the situation, someone analyzing an algorithm might count different things, such as instructions, comparisons, or element swaps. We will generally think in terms of instructions, but one of the benefits of using $O()$ is that in the end, our specific choice probably won't affect the result.

Best, Worst, and Average Case

Algorithms may perform very differently on different data sets. Thus, we may be interested in any of these:

Best-case performance assumes that the data is arranged in such a way that the algorithm performs optimally.

Worst-case performance assumes that the data is arranged as badly as possible for this particular algorithm.

Average-case performance averages over all possible data sets.

Normally, best-case performance is not very helpful because it is more about luck than anything else.

Because of the inequality in its official definition ($g(n) \leq c\,f(n)$), big-O notation is ideally suited for worst-case analysis. Saying that an algorithm is $O(n)$ means that its performance is no worse than linear.

Performance Measurement

A useful counterpart to determining time complexity is to measure an algorithm's performance on an actual machine. **Performance measurement** takes an empirical approach to the question of algorithm performance by measuring the time it takes for code to execute on a particular machine.

Execution time can be measured in Java with the `currentTimeMillis()` function given in Table 2.2. Because it is a static method from another class (the `System` class), this method must be called with a "`System.`" prefix (see page 30).

TABLE 2.2:   System Timing

`static long currentTimeMillis()`
Current time in milliseconds.

The idea is to check the time on a clock immediately before and immediately after running the algorithm:

```
long start = System.currentTimeMillis();
# execute algorithm here
long elapsed = System.currentTimeMillis() - start;
```

Be careful to only time the algorithm in question. Given the speed of current computers, it is common to loop over several executions and then calculate the average time.

An advantage of this approach is that it can be a reality check for an abstract analysis. If your analysis is correct, then real performance times should confirm it. However, beware of machine factors that may impact actual performance, especially caches.

An advantage of $O()$ analysis is that it does not depend on the particular compiler, machine, or in some cases even language, used to code the algorithm. For example, **pseudocode**, which describes algorithms in code-like steps, can often be usefully analyzed using big-O notation.

In the next two sections, we practice algorithm analysis (and using Java arrays) with a pair of examples: insertion sort and binary search.

Exercises

1. Classify these mathematical functions according to their $O()$:

   (a) $13n$

   (b) $n^2 + n + 100$

   (c) 1024

   (d) $\dfrac{n}{100} + 10000$

   (e) $n^2 + n^3$

   (f) $n^2 + n \log n + n$

   (g) $n^{3/2} + n$

2. Determine the $O()$ performance of each of these segments of code. Explain your answers.

   (a)
   ```
   for (int i = 0; i < n; i++) {
       count++;
   }
   ```

   (b)
   ```
   for (int i = 0; i < 1000; i++) {
       count++;
   }
   ```

   (c)
   ```
   for (int i = 0; i < n; i++) {
       for (int j = 0; j < n; j++) {
           count++;
       }
   }
   ```

   (d)
   ```
   for (int i = 0; i < n; i++) {
       for (int j = i; j < n; j++) {
           count++;
       }
   }
   ```

   (e)
   ```
   for (int i = 0; i < n; i++) {
       for (int j = 0; j < 1000; j++) {
           count++;
       }
   }
   ```

3. Explain what it means for an expression to be $O(1)$.

4. Explain what it means to say that an array provides $O(1)$ access to any element in it.

5. Determine the $O()$ performance of the `factorial()` function in Listing 1.1 in terms of its parameter n. Explain your answer.

6. Determine the $O()$ best-case, worst-case, and average performance of the linearSearch() function in Listing 1.3. Explain your answers.

7. Determine the $O()$ best-case, worst-case, and average performance of the sum() function from Exercise 9 in Section 1.3. Explain your answers.

8. Determine the $O()$ best-case, worst-case, and average performance of the max() function from Exercise 10 in Section 1.3. Explain your answers.

9. Determine the $O()$ best-case, worst-case, and average performance of the min() function from Exercise 11 in Section 1.3. Explain your answers.

2.2 Sorting: Insertion Sort

Sorting algorithms are an important family of computational algorithms. An array of items named data with length n is called **sorted** if for all $i < n - 1$,

    data[i] <= data[i+1]

Such an array is called **nondecreasing**; **nonincreasing** is defined similarly.

Insertion Sort

Insertion sort is often used by card players to sort their hands. The idea is to view the data in two sections: the left section is a sorted subset of the data, while the right section contains items that remain to be sorted. At each step of the algorithm, we take the next element from the right section and insert it into the left, in such a way that the left section remains sorted.

Example

Given the array

0	1	2	3	4	5	6	7	8
3	14	7	22	45	12	19	42	6

begin by viewing the first element as sorted (on its own):

0	1	2	3	4	5	6	7	8
3	14	7	22	45	12	19	42	6

Then take the next element in the right half and insert it in its proper, sorted place on the left. In this case, 14 can stay where it is because $3 < 14$.

0	1		2	3	4	5	6	7	8
3	14		7	22	45	12	19	42	6

The next element, 7, needs to be inserted between 3 and 14. This requires saving the 7, copying 14 into slot 2, and then copying 7 into slot 1:

0	1	2		3	4	5	6	7	8
3	7	14		22	45	12	19	42	6

This process continues until there are no elements on the right, in which case the array is sorted. Inserting one element at a time in this way is an example of an **incremental strategy**.

Implementation

Listing 2.1 implements insertion sort for integer arrays in Java. Pay close attention to the inner loop, which shifts elements in the array so that the key can be inserted at its proper location.

Listing 2.1: Insertion Sort

```java
1   // Add to class ArrayFunctions
2   public static void insertionSort(int[] data) {
3     for (int i = 1; i < data.length; i++) {
4       int key = data[i];
5       int j = i - 1;
6       while (j >= 0 && key < data[j]) {
7         data[j + 1] = data[j];
8         j--;
9       }
10      data[j + 1] = key;
11    }
12  }
```

Worst-Case Analysis

Insertion sort has a while-loop inside a for-loop. That makes it difficult to analyze because we need to determine how many times the inner loop runs.

The outer loop on line 3 runs $n - 1$ times on an array of length n. The inner loop on line 6 depends on the data. In the worst case, the array will be in reverse sorted order, and every item will have to move the largest possible distance. In that case, item i will cause the while-loop to run i times. Therefore, the total number of inner loop iterations in the worst case is:

$$1 + 2 + \cdots + (n-1)$$

The question is: what is this expression in terms of $O()$? Sums like this occur frequently in algorithm analysis, so it is good to know how to handle them.

A heuristic for finding this sum is to pair its elements from the outside in:

$$1 + 2 + 3 + \cdots + (n-3) + (n-2) + (n-1)$$

$$3 + (n-3) = n$$
$$2 + (n-2) = n$$
$$1 + (n-1) = n$$

Each pair sums to n, and there are $\dfrac{n-1}{2}$ pairs, so the total is

$$\text{Pair sum} \times \text{Number of pairs} = \frac{n(n-1)}{2}$$

which is $O(n^2)$.

Generating Random Data

It would be difficult to test insertion sort (and many other methods) without the ability to generate random test data. Listing 2.2 shows how to use one of the Java libraries to generate random values. It creates a generator object, and then asks the generator for each value to put in the array.

Listing 2.2: Random Data

```java
// Add to class ArrayFunctions:

import java.util.Random;

public static void randomFill(int[] data, int max) {
    Random gen = new Random();
    for (int i = 0; i < data.length; i++) {
        data[i] = gen.nextInt(max);
    }
}
```

The generator is created in line 6, and its nextInt() method is called in line 8 to fill in the next array value.

Using Java Libraries

To use Java libraries, it is standard practice to **import** the library classes that you need at the top of your file:

```
import library.name;
```

In Listing 2.2, the name of the library being used is `java.util`, and the name being imported on line 3 is the class `Random`. The full name of the class is `java.util.Random`, and in fact, this name can be used at any time without an **import**. As mentioned, though, it is a good idea to import such names at the top of your class definition files.

A few instance methods from the `java.util.Random` class are listed in Table 2.3. Consult the API documentation for further details and a complete list.

TABLE 2.3: java.util.Random Methods

double `nextDouble()`
Double chosen uniformly from doubles in $[0, 1)$.
double `nextGaussian()`
Double chosen from normal distribution with mean 0.0 and standard deviation 1.0.
int `nextInt()`
Integer chosen uniformly from all possible **int** values.
int `nextInt(`**int** n`)`
Integer chosen uniformly from the set $\{0, 1, 2, \ldots, n - 1\}$.

Exercises

1. Finish the example of insertion sort begun on page 38.

2. Explain why the worst case for insertion sort is with reverse-sorted data.

3. Show the operation of insertion sort on these arrays:

    (a) {31, 7, 2, 34, 10, 5, 40, 22}

    (b) {29, 12, 48, 41, 19, 6, 25, 33}

    (c) {21, 31, 39, 22, 18, 38, 25, 6}

    (d) {24, 43, 42, 33, 37, 31, 40, 8}

4. Use Listing 2.1 to:

   (a) Explain why the for-loop in line 3 starts at 1 instead of 0.

   (b) Explain why the array access `data[j]` in line 6 is guaranteed to be valid.

   (c) Explain why the loop in line 6 runs at most `i` times. (Hint: look at what happens to `j`.)

5. Analyze the best-case performance of insertion sort using $O()$ notation. Explain your work, and describe the data that leads to best-case performance.

6. Analyze the average-case performance of insertion sort using $O()$ notation by considering the average number of times the inner while-loop will run. Explain your work.

7. Add `insertionSort()` from Listing 2.1 to the `ArrayFunctions` class. Include code in `main()` to test insertion sort on a short array of integers.

8. Add `randomFill()` from Listing 2.2 to the `ArrayFunctions` class. Include code in `main()` to test insertion sort on an array of 100 random integers.

9. Write an `isSorted(int[] data)` method in the `ArrayFunctions` class that returns true if the array is sorted and false otherwise. Use it in `main()` to test insertion sort.

10. Write an overloaded version of `randomFill(int[] data)` that does not limit the size of the random integers it fills the array with. Test your method in `main()`.

11. Measure the performance of insertion sort on random integers. Begin with an array of size $n = 100$ and double the array size up to 200,000. Use the version of `randomFill()` from Exercise 10. Report your results and observations.

12. Test these methods from the exercises of Section 1.3 on random integer arrays of size 100:

    (a) `sum()` method from Exercise 9.

    (b) `max()` method from Exercise 10.

    (c) `min()` method from Exercise 11.

    (d) `display()` method from Exercise 12.

13. **Selection sort** is another natural sorting method that begins by finding the smallest element in the array and swapping it with the first. Then the second-smallest element is found and swapped with the second, and so on until the array is sorted.

   (a) Show the operation of selection sort on the arrays in Exercise 3.
   (b) Implement a `selectionSort()` method in the `ArrayFunctions` class with tests in `main()`.
   (c) Determine the $O()$ time complexity of selection sort.
   (d) Measure the performance of selection sort as in Exercise 11.
   (e) Compare selection sort with insertion sort in terms of how sensitive the performance of each is to the order of the original data.

2.3   Searching: Binary Search

We have already seen one search algorithm: linear search in Section 1.3. It works by checking each item, one by one, until it finds the search target. Linear search has $O(n)$ performance, which is reasonable if the list of elements is not too long.

However, if an array is sorted, then a more efficient **divide-and-conquer** strategy can be used. It compares the target with the middle item in the array. Because the array is sorted, if the target is smaller than the middle element, then the entire upper half of the array can be eliminated from consideration. Similarly, the lower half is eliminated if the target was larger than the middle item. Repeating this process leads to the algorithm known as **binary search**.

Implementation

Binary search is delicate, in the sense that it can be difficult to write correctly for all cases.[1] The implementation in Listing 2.3 ensures that if the target is in the array, then it must be trapped between two indices `left` and `right`:

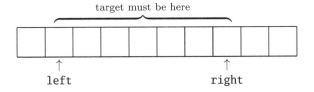

---

[1]See Column 4 from Bentley [2]. An interesting blog post by Bloch [3] points out an error even in that version. The same error is in Listing 2.3 because this code is intended for beginners, but interested readers are encouraged to find and fix it.

At each stage, the item in the middle of this range is checked, and the algorithm stops if `left` is ever greater than `right`.

Listing 2.3: Binary Search

```java
 1  // Add to class ArrayFunctions
 2  public static int binarySearch(int[] data, int target) {
 3      int left = 0;
 4      int right = data.length - 1;
 5      while (left <= right) {
 6          int mid = (left + right) / 2;
 7          if (target == data[mid]) {
 8              return mid;
 9          } else if (target < data[mid]) {
10              right = mid - 1;
11          } else {
12              left = mid + 1;
13          }
14      }
15      return -1;
16  }
```

Sample Trace

The trickiest part of binary search is updating either `left` or `right` when the target is not at `data[mid]`. **Tracing** intricate code by hand is a good way to understand how it works. For example, suppose we search for 15 in this sorted array:

0	1	2	3	4	5	6	7	8
3	6	7	12	14	19	22	42	45

target: 15

Then at the start of the algorithm, we have:

    left: 0   right: 8   mid: 4

Because `target` is greater than `data[mid]`, `left` is changed to `mid + 1`, which is 5, and the next `mid` is calculated (remember that integer division truncates toward zero):

    left: 5   right: 8   mid: 6

In this case, `target` is less than `data[mid]`, so `right` is changed to 5:

    left: 5   right: 5   mid: 5

The last step may be unexpected. Since `target < data[mid]`, `right` is set to `mid - 1`:

```
left: 5                    right: 4
```

and this stops the binary search.

Analysis of Binary Search

Listing 2.3 looks difficult to analyze because of the unknown number of repetitions in the loop on line 5. However, we can simplify a worst-case analysis by assuming that the array size n is a power of 2.

If $n = 2^k$, then in the worst case, binary search will divide the array in half approximately k times. To see why, watch the approximate size of the region `data[left...right]`, where the target must be, after each iteration:

	Size of data[left..right]
After 1 iteration	2^{k-1}
After 2 iterations	2^{k-2}
After 3 iterations	2^{k-3}
...	
After k iterations	$2^{k-k} = 2^0 = 1$

This means that in the worst case, the loop runs approximately k times, and so, binary search is $O(k)$ for $n = 2^k$. To obtain an expression in terms of n, we use the fact:

$$\text{If } n = 2^k, \text{ then } k = \log_2 n$$

Thus, binary search is a $O(\log n)$ algorithm.

Exercises

1. Show how binary search works on the array

   ```
   {2, 5, 7, 10, 22, 31, 34, 40}
   ```

   when searching for each of the elements below. Show the values of `left`, `right`, and `mid` as they change.

   (a) 31   (d) 50

   (b) 2   (e) 1

   (c) 17   (f) 34

2. Show how binary search works on the array

    {10, 13, 24, 36, 37, 41, 44, 66, 86, 100}

   when searching for each of the elements below. Show the values of left, right, and mid as they change.

   (a) 25   (d) 99

   (b) 24   (e) 100

   (c) 86   (f) 8

3. Explain why best-case performance is not helpful for search methods like linear search or binary search.

4. Determine which is faster on unsorted data: linear search, or insertion sort followed by binary search. Assume worst case.

5. Add binarySearch() from Listing 2.3 to the ArrayFunctions class, and write code in main() to call binary search on a short, sorted array of integers. Use an array initializer.

6. Test binarySearch() on a sorted array of 100 random integers. Use insertionSort() from Section 2.2 to sort the random data.

7. It may seem as though lines 10 and 12 in Listing 2.3 are unnecessary optimizations. Replace those lines of code with the simpler alternatives:

    ```
    right = mid;
    ```

   and

    ```
    left = mid;
    ```

   Describe the result.

Chapter 3

Integer Stacks

We now turn to the study of data structures. A **data structure** is just a way of storing data, along with a set of operations for manipulating that data. If you have used anything like a list, map, array, or dictionary, then you already have some experience with data structures. Our purpose here is to begin a systematic study of common data structures, focusing on their use, implementation, and performance. More advanced features of Java will also be described as we need them along the way.

We begin with a simple data structure known as a stack.

3.1 Stack Interface

A **stack** is an abstraction of a vertical stack of physical objects. For example, imagine a tall stack of books. It is hard to pull a book out from the middle or bottom of the stack. However, it is easy to add a new book to the top of the stack; this is called **pushing** onto the stack. And it is similarly easy to remove the top book from the stack, which is called **popping** the top of the stack.

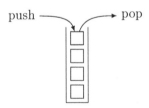

Stacks *only* allow these two ways of adding and removing items: pushing onto the top and popping from the top. Stacks are also known as a **last-in, first-out (LIFO)** data structures, because the last element inserted will be the first one removed.

Table 3.1 gives a complete list of Integer Stack methods. With it, you can begin writing code using stacks. Types described in this way, via a list of public methods, are known as abstract data types.

TABLE 3.1:   Integer Stack ADT

boolean isEmpty()
Returns true if stack has no elements in it.
int peek()
Returns item at top of stack without removing it.
int pop()
Returns and removes item from top of stack.
void push(**int** item)
Adds item to top of stack.
int size()
Number of elements in stack.

Abstract Data Types

An **abstract data type (ADT)** defines the operations of a data type, also known as its **interface**, without specifying the implementation of those operations. This creates an **abstraction**, where interface is separated from implementation in order to hide the details of the implementation:

Interface	Implementation
What an object can do	How it will be done

Abstract data types are usually described by their **application programming interface (API)**, which is a list of the public methods available for that type. Table 3.1 gives the API of the Integer Stack ADT.

In fact, you already have experience with this idea. Think about the Java objects you have used so far, such as String, StringBuilder, or Random objects. You did not need to know how any of these objects were implemented in order to use them, as long as you knew their public interfaces. And the Java API documentation [10] provides exactly this information—the public interface or API—for all Java library classes.

Thus, all Java classes support abstract data types via their public methods and documentation. But there is more.

Java Interfaces

Java interfaces allow us to express ADTs directly in code. Java interfaces contain method declarations, describing what an object can do, but no implementation code. It is up to classes to implement the interface and provide code for implementation.

The syntax to define an interface simply uses the keyword **interface** instead of **class**:

```
public interface InterfaceName {
    // field declarations (must be static final)
    // method declarations
}
```

All declarations inside an interface are assumed to be public. Interfaces may also define nested classes and other interfaces, but those techniques are beyond the scope of this text.

Listing 3.1 shows how to write the Integer Stack ADT from Table 3.1 as a Java interface.

Listing 3.1: Integer Stack Interface

```
1  public interface IntStack {
2      boolean isEmpty();
3      int peek();
4      int pop();
5      void push(int item);
6      int size();
7  }
```

Normally, documentation comments would be included in Listing 3.1 to explain what each method is intended to do. Then javadoc would convert those comments into a form like Table 3.1. We omit documentation comments to help focus attention on the code.

Interface Types

Interfaces define new data types in Java, much like classes. Interface types fall into the family of reference types:

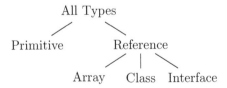

One of the main purposes of an interface like `IntStack` is to use the type to declare variables. The declaration

```
IntStack s;
```

allows s to call any of the methods in `IntStack`. However, because it contains no implementation code, an interface may *not* be used to create objects. That means s cannot call any methods until it refers to an object that implements the interface.

Classes Implement Interfaces

Classes that **implement** an interface must provide all of the methods listed in the interface, and those methods must be declared public. The syntax to declare that a class implements an interface uses the **implements** keyword:

```
public class ClassName implements InterfaceName { ... }
```

A class may implement more than one interface by separating the interface names with commas.

Once a class implements an interface, it can be used to create objects that interface variables refer to. For example, if the class `IntArrayStack` implements `IntStack`, then this is valid:

```
IntStack s = new IntArrayStack();
s.push(10);
System.out.println(s.pop());        // output 10
```

If the `IntArrayStack` class has public members other than those in the interface, the variable s cannot use them because it was declared with type `IntStack`.

Programming to Interfaces

In general, try to declare variables with interface types rather than class types. This technique, known as **programming to the interface**, allows greater flexibility because it does not tie your code down to particular implementations. If you declare variables with interface types, then changing to a new implementation for that variable reference is usually just a matter of calling a different constructor, such as:

```
IntStack s = new IntLinkedStack();
```

No subsequent code using s should have to change because it is already using the interface.

Exercises

1. Show the results of these operations on an initially empty IntStack s.
   Draw the stack contents after each operation, making clear where the
   top is, and indicate the return value of all non-void methods.

   (a) s.push(5)   (b) s.push(1)   (c) s.push(10)
       s.push(8)   s.pop()   s.push(20)
       s.peek()   s.push(2)   s.push(30)
       s.push(3)   s.peek()   s.peek()
       s.pop()   s.pop()   s.push(40)
       s.push(10)   s.push(3)   s.push(50)
       s.size()   s.size()   s.size()
       s.push(4)   s.pop()   s.pop()
       s.pop()   s.push(4)   s.pop()
       s.pop()   s.pop()   s.pop()
       s.isEmpty()   s.isEmpty()   s.isEmpty()

2. Decide whether or not a stack would be an appropriate data structure
   for each of these types of task lists. Assume that when a new task
   arrives, it would be pushed, and when a task is chosen to work on, it
   would be popped. Explain your answers.

   (a) Tasks that may need to be done in any order.

   (b) Tasks where the next one to work on is always the one that has
       been waiting the longest.

   (c) Tasks where the next one to work on is always the most recently
       received.

   (d) Tasks that need to be done in the order they are received.

   (e) Tasks that may need to be shuffled or sorted.

3. Explain the difference between a class and an interface in your own
   words.

4. Suppose IntArrayStack and IntLinkedStack implement the IntStack
   interface. Write one Java statement to do each of these tasks:

   (a) Declare s to be an IntStack.

   (b) Set s from part (a) to refer to a new IntLinkedStack.

   (c) Declare t to be an IntStack referring to a new IntArrayStack.

5. Suppose s refers to an `IntStack` object. Write Java code to accomplish these tasks:

   (a) Add the value 100 to the top of s.

   (b) Remove and print the value at the top of s.

   (c) Print the value at the top of s without removing it.

   (d) Remove and print every item in s until it is empty.

6. Write a Java interface named `Runnable` with one void method `run` that takes no parameters.

7. Write a Java interface named `Queue` with two methods: a void method `enqueue` that takes an **int** parameter, and a method `dequeue` with no parameters that returns an **int**.

8. Suppose the classes `LinkedQueue` and `ArrayQueue` both implement the `Queue` interface from Exercise 7. Write single Java statements for each of these tasks:

   (a) Declare a queue `q1` that refers to a new array queue.

   (b) Declare a queue `q2` that refers to a new linked queue.

   (c) Call the `enqueue` method on `q1` with the value 19.

   (d) Store the result of calling `dequeue` on `q2` in a variable named `result`.

3.2   Array Implementation

In this section and the next, we write two very different implementations of the `IntStack` interface. The first uses an array.

Suppose we plan to use an array to store the contents of a stack, and that we begin with an empty stack and push the integers 10, 20, and 30 (in that order). Then there are two natural options: store the top element at the front of the array or in the back.

Top at front

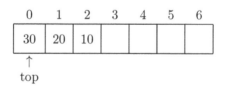

Top in back

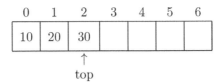

top

To get a feel for how these two options work, suppose the next operation is to push 40.

Top at front The existing items have to shift to make room for the new element:

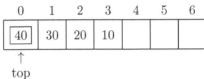

top

Shifting (and therefore pushing) takes $O(n)$ time, because each element has to move one space to the right. Popping is also $O(n)$ because after removing the top, the remaining elements have to each shift one space left. Notice that it may also be difficult to know when the stack is empty.

Top in back In this case, 40 can go directly to slot 3:

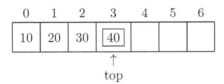

top

This takes $O(1)$ time, as does popping, because no shifting is required. The stack is empty whenever `top` is -1.

Given its superior performance, Listing 3.2 begins the implementation of an integer stack using the second approach. As is, it will not compile because the class does not completely implement the `IntStack` interface. You will be asked to complete the `IntArrayStack` class in the exercises.

Throwing Exceptions

Line 18 of Listing 3.2 demonstrates how to handle an attempt to pop from an empty stack. Without that if-statement, trying to pop from an empty stack would cause an `ArrayIndexOutOfBoundsException`. However, someone using the stack should know that they cannot pop from an empty stack, and the `isEmpty()` method is provided to help them check. There is also no way to "fix" this error—if the stack is empty, there is nothing that can be popped.

Listing 3.2: Integer Array Stack

```java
import java.util.EmptyStackException;

public class IntArrayStack implements IntStack {
    private int top = -1;
    private int[] data;
    private static final int DEFAULT_CAPACITY = 10;

    public IntArrayStack() {
        data = new int[DEFAULT_CAPACITY];
    }

    public void push(int item) {
        if (top == data.length - 1) resize(2 * data.length);
        data[++top] = item;
    }

    public int pop() {
        if (isEmpty()) throw new EmptyStackException();
        return data[top--];
    }

    private void resize(int newCapacity) {
        int[] newData = new int[newCapacity];
        for (int i = 0; i <= top; i++) {
            newData[i] = data[i];
        }
        data = newData;
    }

    public static void main(String[] args) {
        IntStack s = new IntArrayStack();
        for (int i = 0; i < 5; i++) {
            s.push(i);
        }
    }
}
```

In situations like this, we pass the problem back to the caller by throwing an exception:

```
throw new ExceptionName();
```

Throwing an exception causes the current method to stop execution immediately, and gives the caller a chance (and the responsibility) to deal with the mistake. In this case, the empty stack exception is more informative than a generic array index exception would have been.

Table 3.2 lists two of the exceptions available in the `java.util` library. Because they come from a library, remember to import these exceptions before using them.

TABLE 3.2: java.util Exceptions

EmptyStackException
Indicates the stack is empty.
NoSuchElementException
Indicates the requested element does not exist.

Resizing the Array

Recall that once an array has been created, its length can never be changed. Therefore, if `push()` is called when the array is full, we need to create a new array, copy the existing stack items into the new array, and then set `data` to refer to the new array. We put this work into a private `resize()` method in Listing 3.2. By doubling the length each time the array becomes full, we minimize the impact of resizing on the performance of `push()`. However, calculating that precise effect is beyond the scope of this text.

Java has built-in system methods for copying arrays that should be used in production code, but we are doing it by hand to practice working with arrays. In addition, the queue array implementation will need its own specialized resizing method.

Exercises

1. Write the Java code to declare an `IntStack` named s that refers to an `IntArrayStack`, and then push the values 10, 20, 30, and 40 onto s.

2. Write the Java code to declare an `IntStack` named `operands` referring to an `IntArrayStack`, and then push the values 17, 0, −12, and 101 onto `operands`.

3. Use Listing 3.2 to:

    (a) Explain why the starting value of top is -1.

    (b) Explain why the DEFAULT_CAPACITY variable is declared **final** and **static**.

    (c) Explain the use of prefix increment in line 14.

    (d) Explain the use of postfix decrement in line 19.

    (e) Explain why, without line 18, a call to pop() on an empty stack would cause an ArrayIndexOutOfBoundsException.

4. Suppose s is a reference to an IntArrayStack. Explain the difference between s.size() and s.data.length.

5. Draw the contents of the s.data array after the main() method of Listing 3.2 has run. Indicate the value of s.top.

6. Finish the IntArrayStack class of Listing 3.2 by adding these methods from the IntStack interface:

    (a) isEmpty()

    (b) peek()   Throw an exception if the stack is empty.

    (c) size()

7. Modify Listing 3.2 to:

    (a) Rewrite the push() method without using the prefix increment operator. Discuss the tradeoffs.

    (b) Rewrite the pop() method without using the postfix decrement operator. Discuss the tradeoffs.

8. Modify the main() method of Listing 3.2 to:

    (a) Pop and print each item in the stack until it is empty.

    (b) Push and pop a large number of items to test the array resizing.

    (c) Test the isEmpty() method both before and after items have been pushed.

    (d) Test the size() method before and after items have been pushed.

9. Explain why each of these IntArrayStack methods is $O(1)$:

    (a) isEmpty()

    (b) peek()

    (c) pop()

    (d) push()   Ignore resizing.

    (e) size()

10. Modify Listing 3.2 to add a second constructor with one parameter specifying the starting capacity of the stack. Modify the existing constructor to call the new constructor using **this()** (see page 29).

11. Modify the **pop()** method of Listing 3.2 to reduce the array length by half if the number of elements in the stack is less than or equal to one-fourth the current length of the array (but do not let the array have a size smaller than 10). Include an output statement in the **resize()** method to report each resizing, and then test your modification by pushing and then popping a large number of elements.

12. Instead of having a **top** instance variable, the **IntArrayStack** class could use a **size** instance variable that stores the number of elements in the stack. Rewrite Listing 3.2 to use a **size** instance variable instead of **top**. (In other words, your code should not use **top** at all; all calculations should be done in terms of **size**.) Discuss the tradeoffs between these two approaches.

3.3 Linked Implementation

Although the array implementation may seem natural, it is not the only way to build a stack. Think again about the reason we didn't keep the top of the stack at the front of the array implementation (pages 52–53): pushing and popping required shifting elements in the array to either make room or close gaps. A linked implementation allows us to insert or remove items at any location without having to shift elements already in place. The tradeoff is that linked structures lack the direct, indexed access to every item that arrays have.

Linked Lists

The simplest linked structure is a **linked list**, a collection of **nodes** in which each node points to the node that follows it. We will use nodes with two fields: **data** to hold whatever it is we are storing in the list, and a **next** reference that refers to the next node in the list:

data next

Nodes link together to form a list:

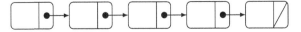

A head pointer is used to point to the front of the list, and the end of the list is indicated by a node with a null reference in its **next** field.

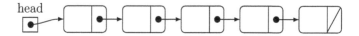

Using a linked list to implement stacks will allow us to keep the top of the stack at the front of the list, where it is easy to add and remove items without having to move any of the other stack contents. Only a few references will need to change.

Nodes

A Node class is used to represent the nodes that link together to form linked lists. A typical Node class has only its two fields, **data** and **next**, and a constructor, as in Listing 3.3.

Listing 3.3: Node

```
1  public class Node {
2      private int data;
3      private Node next;
4
5      public Node(int data, Node next) {
6          this.data = data;
7          this.next = next;
8      }
9  }
```

To implement a stack, we only need two linked list operations: insert and remove from the front of the list.

Insertion at Front

Inserting a new item at the front of a linked list takes three steps:

1. Create a new node:

   Node p = **new** Node(item, **null**);

2. Set the **next** field of the new node to point to the existing first node:

```
p.next = head;
```

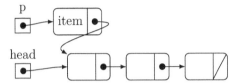

3. Update **head** to point to the new node:

```
head = p;
```

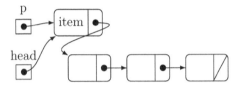

That's it. It helps to see these steps in sequence, but we can actually combine them all into one line of code. The first two steps can be done in one by providing the **next** field of the new node to the constructor:

```
Node p = new Node(item, head);
```

The third step changed **head** to point to the new node, so we can bypass that as a separate step by avoiding the name **p** altogether and just setting **head** to point to the new node. Thus, inserting a new node at the front of a linked list boils down to one line of code:

```
head = new Node(item, head);
```

Assignment statements always evaluate the expression on the right first and then assign that value to the variable on the left. Thus, the new node is created using the old value of **head** as its **next** field, and then **head** is assigned to the new node.

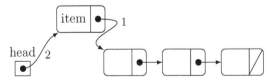

Deletion at Front

Deleting a node at the front of a linked list is also one step. Just update **head** to point to the second node:

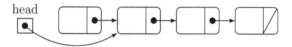

The second node is referenced by **head.next**, and so the code is:

```
head = head.next;
```

If you need to recover data from the node that was deleted, then that has to happen before the **head** pointer is changed.

Other linked list operations will be discussed later as we need them.

Linked Implementation

Listing 3.4 contains the core operations of a linked stack. The exercises ask you to finish it.

Nested Classes

Because code outside the **IntLinkedStack** class has no reason to know about nodes, the **Node** class is defined inside the body of **IntLinkedStack**. That makes the **Node** class a **nested** class of **IntLinkedStack**. Nested classes are considered members of their containing class, just like fields and methods. Usually, nested classes are also declared private because they are implementation details of the containing class, similar to fields. Nested classes are always visible to their containing class (and vice versa), even when declared private.

A **static nested class** has no access to the instance variables of the containing class. In Listing 3.4, the only instance variable of the stack is **top**, and there is no reason for every node to have access to the top of the stack. Most of our linked structures will use static nested node classes, but we will see an example of a non-static nested class in Chapter 6.

Null Pointer Exceptions

Recall (see page 22) that attempting to access a field or method of a null reference generates a **NullPointerException**. Linked list code is particularly susceptible to null pointer exceptions because of how often expressions like **p.next** and **p.next.next** are used.

Listing 3.4: Integer Linked Stack

```java
public class IntLinkedStack implements IntStack {
    private Node top;

    public void push(int item) {
        top = new Node(item, top);
    }

    public int pop() {
        int result = top.data;
        top = top.next;
        return result;
    }

    private class Node {
        private int data;
        private Node next;

        private Node(int data, Node next) {
            this.data = data;
            this.next = next;
        }
    }
}
```

Thus, every time you access a field or method using dot notation such as

p.next   — *Whatever is here cannot be null*

ask yourself how you know for sure that the reference is not null.

Exercises

1. Look again at Exercise 1 from Section 3.1. Do your answers to it depend on whether the list uses an array or a linked implementation? Explain why or why not.

2. Draw a linked list containing the elements {5, 10, 4, 8}.

3. Draw the linked list that results from running this segment of code:

```java
Node head = new Node(1, null);
head = new Node(2, head);
head = new Node(3, head);
```

4. Write Java statements using nodes, not stack operations, to create a linked list with contents {7, 4, 18}.

5. Suppose p references the last node in a linked list, so that p.next is null. Which of the following references (if any) will generate a null pointer exception? Explain your answers.

   (a) p

   (b) p.next

   (c) p.next.next

6. Modify Listing 3.4 so that it contains the same main() method as Listing 3.2, with the constructor call replaced by a call to IntLinkedStack(). Draw the contents of the linked list that results, including s.top.

7. Explain why, as written in Listing 3.4, trying to pop from an empty stack would cause a null pointer exception.

8. Modify Listing 3.4 to throw an EmptyStackException if pop() is called on an empty stack.

9. Finish the IntLinkedStack class of Listing 3.4 by adding these methods from the IntStack interface:

   (a) isEmpty()

   (b) peek()   Throw an exception if the stack is empty.

   (c) size()   Add a size instance variable and update it where necessary.

10. Modify the main() method of the IntLinkedStack class to:

    (a) Pop and print each item in the stack until it is empty.

    (b) Push and pop a large number of items.

    (c) Test the peek() method.

    (d) Test the size() method.

    (e) Test the isEmpty() method.

11. Explain why a constructor with a capacity parameter (as described for the IntArrayStack class in Exercise 10 of Section 3.2) is not appropriate for the IntLinkedStack class.

12. Explain why each of these `IntLinkedStack` methods is $O(1)$:

   (a) `isEmpty()`

   (b) `peek()`

   (c) `pop()`

   (d) `push()`

   (e) `size()`   Is the performance the same if there is no `size` instance variable? Explain.

Chapter 4

Generic Stacks

The previous chapter built an interface and two implementations for an integer stack. Suppose now that we need a stack to hold strings. (We are about to need one, in fact, in Section 4.4.) With copy-paste and some careful editing, we could create a `StringStack` interface and two corresponding string implementations. But then what if we need a stack to hold some other type? You can see the problem: continuing to write new classes and interfaces for every element type creates a lot of code that is essentially the same. What we need is a **type variable**, so that each time we make a new stack, we can specify the type that will be stored in it. This is known as **generic programming**.

4.1 Generic Types

Type Parameters

A **generic type** (interface or class) is defined by including one or more **type parameters** in angle brackets at the end of the name of the interface or class:

```
public interface InterfaceName<TypeParameter, ...> { ... }
```

or

```
public class ClassName<TypeParameter, ...> { ... }
```

Type parameters are usually single capital letters, such as those in Table 4.1.

TABLE 4.1: Type Parameters

E	Element type in a collection
K	Key in a key-value pair
V	Value in a key-value pair
T, U, S	All-purpose

For example,

```
public interface Map<K, V> { ... }
```

defines a generic `Map` interface with two type parameters, K and V.

Generic Stack ADT

Generic type parameters allow us to define the generic Stack ADT in Table 4.2, corresponding to the Integer Stack ADT in Table 3.1. Most data structures use the E type parameter for elements in a collection.

TABLE 4.2:   Stack ADT

boolean isEmpty()   Returns true if stack has no elements in it.
E peek()   Returns item at top of stack without removing it.
E pop()   Returns and removes item from top of stack.
void push(E item)   Adds item to top of stack.
int size()   Number of elements in stack.

Type Parameters Inside Interface and Class Definitions

Inside a generic class or interface definition, the type parameter and the generic type itself may be used for declarations as if they were regular types. (There is an important limitation with arrays that will be addressed in the next section.) For example, Listing 4.1 contains code for a generic Node class.

Listing 4.1: Generic Node

```
1  public class Node<T> {
2     private T data;
3     private Node<T> next;
4
5     public Node(T data, Node<T> next) {
6        this.data = data;
7        this.next = next;
8     }
9  }
```

Both T and Node<T> are used as types to declare variables inside the generic Node class.

Type Arguments

In order to use a generic interface or class, a reference type must be specified as a **type argument** for the type parameter. The type argument is then used as the value of the type parameter.

For example, using the `Node` class above, we can write:

```
Node<String> p;
```

to declare a reference to a Node that stores a string. In this case, `String` is supplied as the type argument for the type parameter T. This is exactly analogous to providing arguments to method parameters, except that types are involved.

Warning: Type arguments must be reference types; they cannot be primitive. To store primitive data in a generic type, the corresponding wrapper class must be used.

Wrapper Classes

Every primitive type has an associated class known as its **wrapper class**, listed in Table 4.3. The idea is that the primitive is "wrapped" in an object of the wrapper class type. Wrapper objects also provide many convenience methods for working with their values; see the Java API [10] for details.

TABLE 4.3: Wrapper Classes

Primitive Type	Wrapper Class
byte	Byte
short	Short
char	Character
int	Integer
long	Long
float	Float
double	Double
boolean	Boolean

Since primitive types cannot be used as generic type arguments, the corresponding wrapper class must be used instead:

```
Node<int> p;      // incorrect -- int is not a reference type
Node<Integer> p;  // correct
```

Although working with primitive types and wrapper classes can be awkward at times, Java attempts to make it as convenient as possible via automatic boxing and unboxing.

Automatic Boxing and Unboxing

If you use a primitive value where a corresponding wrapper object is expected, Java will automatically **box** the primitive in an object; similarly, it will automatically **unbox** a wrapped object to a primitive when necessary.

For example, if p refers to a Node<Integer> and x has type **int**, then Java automatically handles the conversions between the Integer p.data and the **int** x:

```
p.data = x;       // x is automatically boxed
x = p.data;       // p.data is automatically unboxed
```

Casting can also be used to explicitly box or unbox.

Creating Generic Objects

Beginning with Java 7, the **diamond operator** <> is used with **new** to create all generic objects:

```
referenceVariable = new ClassName<>(arguments);
```

In other words, the type argument is not specified when creating objects, only when declaring the type of the reference variable.

For example, the following creates an integer node referenced by **p**:

```
Node<Integer> p = new Node<>(item, null);
```

Earlier versions of Java require that you also specify the type argument (in this case, Integer) inside the diamond.

Exercises

1. Write a generic interface Queue<E> with two methods: a void method **enqueue** that takes a parameter of type E, and a method **dequeue** with no parameters that returns an element of type E.

2. Suppose ArrayQueue<E> implements Queue<E> from Exercise 1. Write a Java statement to declare and create an array queue q that contains strings. Use the interface to declare the type.

3. Write a generic interface Map<K, V> with two methods: a void method **put** that takes two parameters of type K and V, and a method **get** that takes a parameter of type K and returns an item of type V.

4. Suppose LinkedMap<K, V> implements Map<K, V> from Exercise 3. Write a Java statement to declare and create a linked map m that uses string keys and integer values. Use the interface to declare the type.

5. Write a generic Java interface, `Stack<E>`, for the generic Stack ADT in Table 4.2.

6. Suppose `ArrayStack<E>` implements the `Stack<E>` interface from Exercise 5.

   (a) Write a Java statement to declare and create a string array stack `operators`. Use the interface to declare the type.

   (b) Write a Java statement to declare and create an integer array stack `operands`. Use the interface to declare the type.

   (c) Write Java code to declare and create an integer array stack `s`, and then push the values 0 through 9 onto the stack.

7. Suppose `s` is declared as a `Stack<Integer>` using the interface from Exercise 5. Determine whether or not there is any automatic boxing or unboxing in this code:

   ```java
   for (int i = 0; i < 100; i++) {
       s.push(i);
   }
   ```

8. Use the generic `Node` class in Listing 4.1 to:

   (a) Declare and create an integer node named **p** containing the value 34.

   (b) Declare and create a string node named **q** containing the value "Java."

   (c) Build a linked list of doubles pointed to by **head** containing the values 3.14, 2.72, and 0.58.

4.2 Generic Stack Implementations

Exercise 5 in the previous section asked you to define the generic stack interface `Stack<E>`. In this section, we create generic versions of the linked list and array stack implementations. The basic idea should be clear:

$$\text{IntLinkedStack} \Rightarrow \text{LinkedStack<E>, using E instead of } \textbf{int}$$
$$\text{IntArrayStack} \Rightarrow \text{ArrayStack<E>, using E instead of } \textbf{int}$$

Not surprisingly, there are a few complications. The linked implementation has just a small one, so we address it first.

Generic Nested Classes

The Node<T> class in Listing 4.1 is ready to be inserted as a nested class in the generic LinkedStack<E> class, as long as its visibility is changed to private. The Node<T> class uses the type parameter T intentionally: when declaring a generic nested class, it is best to use a *different* type parameter than whatever was used in the containing class.[1]

Thus, the LinkedStack<E> class should begin as in Listing 4.2. The type parameter E should be used for defining all the methods in the LinkedStack<E> class—T is only used to define the static Node class. The exercises ask you to complete the LinkedStack<E> implementation.

Listing 4.2: Generic Linked Stack

```
1   public class LinkedStack<E> implements Stack<E> {
2       private Node<E> top;
3
4       // use E here
5
6       private static class Node<T> {
7           private T data;
8           private Node<T> next;
9
10          private Node(T data, Node<T> next) {
11              this.data = data;
12              this.next = next;
13          }
14      }
15  }
```

There are a few more difficulties in making a generic ArrayStack<E>. The main issue is due to a limitation of generic arrays.

Generic Arrays

The data array in an IntArrayStack has type **int**[ ] because we store integers in the stack. To make a generic array stack, the data array will need to store values of type E, the generic type parameter.

Generic types may be used to declare array types:

    private E[ ] data;

---

[1]The reasons are subtle; see Puzzle 89 from Bloch and Gafter [5].

However, generic types may *not* be used to create arrays:

data = **new**(E)LENGTH];
    ↖ *No generic types allowed here*

The reason creating generic arrays is prohibited is that the compiler needs to know the array element type E in order to generate the correct code, but the value of E will not be specified until each array stack is created.

The simplest solution is to create an Object array and cast it to the generic type:

```
E[] arrayRef = (E[]) new Object[LENGTH];
```

The cast (see page 3) tells the compiler to treat the array as if it were of type E[]. Because we plan to only push and pop elements of type E, this will not be a problem. The nature of the Object class will be explained later in Section 5.3; for now, think of it as a type that can be any object at all.

Once the array has been created and cast in this way, it can be treated as if it has type E[].

Unchecked Casts

Unfortunately, the compiler cannot verify that casting an Object[] array to a generic E[] array is safe, and so it will produce an **unchecked cast** warning. (If the warning does not specify where the unchecked cast occurred, recompile with the -Xlint:unchecked flag.)

Rather than ignoring the warning, a better solution is to squelch it by informing the compiler that—in this case—we know what we are doing by putting the following **annotation** above the beginning of the method where the cast is used:

```
@SuppressWarnings("unchecked")
```

Of course, this annotation should be used sparingly.

Obsolete References

Finally, because generic type arguments must be reference types, a new issue arises from storing object references in the array instead of primitive types. Although it is not serious, the same problem can occur with any data structure that stores references in an array. Attending to it now will help build your intuition about the differences between primitive and object types.

Consider what an array stack with initial capacity of 7 looks like if we push the integers 0 through 6:

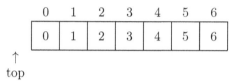

and then pop until the stack is empty (when the top is −1):

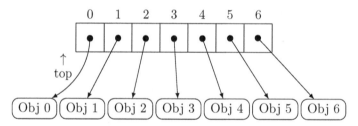

We did not delete any of the items that were popped because there was no reason to: no additional memory is used, and the stack does not allow any access to items past the top.

The situation is different, however, with object references. If we push and then pop the same number of objects, then their references will still be active even when the stack is back to being empty:

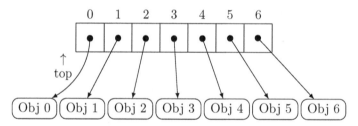

The stack prevents accessing any of those objects, but the objects' memory cannot be garbage collected, because it is still pointed to by active references. As noted on page 23 (and Item 6 from Bloch [4]), the solution is to set those references to null once they are no longer needed.

However, do not get carried away setting all of your references to null; this is a relatively rare problem. And although setting popped entries to null is good practice, do not rely on it to determine which array entries are being used for the stack. The **top** instance variable should always be used to decide where the top of the stack is.

Generic Array Stack

Putting these pieces together results in Listing 4.3, a start to the generic **ArrayStack<E>** class. You will be asked to write the rest of the **ArrayStack<E>** class in the exercises.

Listing 4.3: Generic Array Stack

```
1   public class ArrayStack<E> implements Stack<E> {
2       private E[] data;
3       private int top = -1;
4       private static final int DEFAULT_CAPACITY = 10;
5
6       @SuppressWarnings("unchecked")
7       public ArrayStack() {
8           data = (E[]) new Object[DEFAULT_CAPACITY];
9       }
10
11      public E pop() {
12          E result = data[top];
13          data[top--] = null;
14          return result;
15      }
16  }
```

Exercises

1. Complete the `LinkedStack<E>` class from Listing 4.2 by writing these methods:

   (a) `isEmpty()`

   (b) `peek()`   Throw an exception if the stack is empty.

   (c) `pop()`   Throw an exception if the stack is empty.

   (d) `push()`

   (e) `size()`   Add a `size` instance variable and update it where necessary.

2. Add a `main()` method to the `LinkedStack<E>` class to test the stack with:

   (a) Integers

   (b) Doubles

   (c) Strings

3. Determine the $O()$ performance of each of these `LinkedStack<E>` methods. Explain your answers.

   (a) `isEmpty()`

   (b) `peek()`

   (c) `pop()`

   (d) `push()`

   (e) `size()`

4. Complete the `ArrayStack<E>` class from Listing 4.3 by writing these methods:

    (a) `isEmpty()`

    (b) `peek()`   Throw an exception if the stack is empty.

    (c) `push()`   Double the length of the array if the array is full.

    (d) `size()`

    (e) Private `resize()`

    (f) A second constructor with a starting capacity parameter. Modify the existing constructor to call the new constructor using **this()**.

    (g) Modify `pop()` to throw an exception if the stack is empty.

    (h) Modify `pop()` to reduce the length of the array by half if the number of elements in the stack is less than or equal to one-fourth the current length.

5. Add a `main()` method to the `ArrayStack<E>` class to test the stack with:

    (a) Integers

    (b) Doubles

    (c) Strings

6. Determine the $O()$ performance of each of these `ArrayStack<E>` methods. Explain your answers.

    (a) `isEmpty()`

    (b) `peek()`

    (c) `pop()`

    (d) `push()`   Ignore resizing.

    (e) `size()`

4.3   Evaluating Expressions: Background

Stacks are particularly useful for evaluating numeric expressions. In fact, if we break the evaluation process into two stages, a stack is the key component in both steps. In order to describe this process, we need some terminology first.

Binary Operations

Arithmetical operations like addition, subtraction, multiplication, and division are called **binary operations** because they each combine exactly two **operands**. While we might think the expression $1 + 2 + 3 + 4$ combines more than two operands, in a programming language such as Java, this expression is actually evaluated two terms at a time, grouping from the left:

$$((1 + 2) + 3) + 4$$

Mathematically, the **associative** properties of addition guarantee that all groupings produce the same result; Java chooses this grouping for addition because it treats addition as **left-to-right associative**. Other operations, such as assignment, are **right-to-left associative**, meaning that this code:

```
x = y = z = 10;
```

is evaluated right-to-left as:

```
x = (y = (z = 10));
```

However, associativity only applies to operators with the same precedence. Operator **precedence** is used first to determine the order of operations when parentheses have not been specified. These follow normal mathematical rules, so multiplication and division have higher precedence than addition and subtraction. Consult the Java documentation for a complete list of precedence and associativity rules.

Thus, the final order of operations for an expression in Java is determined by applying each of these in order:

1. Parentheses

2. Operator precedence

3. Associativity among operators with the same precedence

Prefix, Infix, and Postfix

Strangely enough, all of the above rules for parentheses, operator precedence, and associativity exist because we prefer writing binary operations in infix form. Two other forms, prefix and postfix, are completely unambiguous and require none of those additional rules.

Consider the simple binary operation "$a + b$." Equivalent prefix and postfix forms are shown in Table 4.4. Our goal is to become familiar with prefix and postfix notations in order to develop an algorithm for evaluating infix expressions using stacks.

TABLE 4.4:   Operator Notations

Prefix	$+\,a\,b$	Operator first
Infix	$a + b$	
Postfix	$a\,b\,+$	Operator last

Example

When translating prefix or postfix expressions by hand, it can be helpful to identify simple components first, with one operator and two operands, and then build up from there.

For example, consider translating this prefix expression to infix:

```
- / + a b c * d + e f
```

Begin by finding the simple prefix components (operator operand operand):

```
- / [+ a b] c * d [+ e f]
```

Now treat the boxes as operands and repeat the process:

```
- [/ [+ a b] c] [* d [+ e f]]
```

At this stage, we are have one (large) simple expression that can be translated to infix:

```
[/ [+ a b] c] - [* d [+ e f]]
```

Work your way inward, translating each box to infix:

```
[[+ a b] / c] - [d * [+ e f]]
```

```
[[a + b] / c] - [d * [e + f]]
```

Notice that when you write the infix expression, some of the boxes must be written with parentheses to convey the correct expression:

```
(a + b) / c - d * (e + f)
```

Evaluating Postfix Algorithm

One can evaluate postfix expressions by hand in a similar way, and Listing 4.4 systemizes that process using a stack to hold the operands and, eventually, the final result. We will assume all operands are integers.

Listing 4.4: Evaluate Postfix (Pseudocode)

```
1   For each term in expression
2     If term is an operator
3       Pop second operand
4       Pop first operand
5       Apply operator to operands and push result
6     Else
7       Push operand onto stack
8   Pop result
```

Example

To see how Listing 4.4 works, consider the postfix expression

$$3\ 5\ 1\ -\ *$$

The first three terms are operands so they are pushed onto the stack:

$$\boxed{\begin{array}{c} 1 \\ \hline 5 \\ \hline 3 \end{array}}$$

The next term is the operator "$-$", so we follow these steps:

Pop second operand	Stack $\rightarrow$ 1
Pop first operand	Stack $\rightarrow$ 5
Apply operator to operands	$5 - 1$
Push result	Stack $\leftarrow$ 4

Notice the order the operands are popped: the second operand comes off the stack first because of the order in which they were pushed. The stack contents are now:

$$\boxed{\begin{array}{c} 4 \\ \hline 3 \end{array}}$$

The last term is "$*$":

Pop second operand	Stack $\rightarrow$ 4
Pop first operand	Stack $\rightarrow$ 3
Apply operator to operands	$3 * 4$
Push result	Stack $\leftarrow$ 12

and then the stack contains the final result, ready to be popped:

$$\boxed{12}$$

Infix to Postfix Algorithm

Listing 4.5 outlines a similar algorithm to translate infix to postfix. It uses a stack to hold operators and is slightly more complex than Listing 4.4 because of the inner loop in lines 3 and 4. This algorithm assumes there are no parentheses in the infix expression.

Listing 4.5: Infix to Postfix (Pseudocode)

```
1  For each term in expression
2      If term is an operator
3          Pop all operators of same or higher precedence
4              and copy each to output
5          Push this operator onto stack
6      Else
7          Copy operand to output
8  Pop remaining operators and copy to output
```

Example

Given the infix expression

    a + b / c - d

we can trace Listing 4.5 by keeping track of the stack and accumulated output. The first term, a, is copied to output. The second, +, is pushed onto the stack, and the third, b, is copied to output.

At this point, we have:

      +   Output: a b

The next term is the operator /. The + on top of the stack does not have the same or higher precedence as this term, so nothing is popped and / is pushed:

      /
      +   Output: a b

The c is copied to output, and then the next term is the operator -:

      /
      +   Output: a b c

In this case, the / on the top of the stack does have the same or higher precedence than the current term, so it is popped and copied to output:

      +   Output: a b c /

The same is true for +, so it is popped and copied to output:

Output: a b c / +

At this point, the stack is empty, so the – is pushed:

Output: a b c / +

Finally, the d is copied to output and any remaining operators on the stack are popped and copied to the final output:

Output: a b c / + d –

The most difficult part of doing this algorithm by hand is keeping track of when to pop operators and when to leave them on the stack.

The combination of Listings 4.4 and 4.5 gives an algorithm for evaluating infix expressions, which we develop code for in the next section.

Exercises

1. Look up the precedence and associativity of these Java operators: &&, ++, <, and ==.

2. Convert these prefix expressions to infix and postfix:

    (a) * + / – a b c d e

    (b) / – a b * + c d e

    (c) + – a * b c / d e

    (d) – a / b * c + d e

3. Convert these postfix expressions to infix and prefix:

    (a) a b c d e * + / –

    (b) a b * c + d e / –

    (c) a b c + / d e – f / –

    (d) a b + c / d – e * f +

4. Convert these infix expressions to prefix and postfix (without using Listing 4.5, since most have parentheses):

    (a) x + y + z * w – v / u

    (b) (x + y + z) * (w – v) / u

    (c) x * y / (z – w + v * u)

    (d) (x – y) * (z + w) / (u + v)

5. Evaluate these prefix expressions:

    (a) + 3 * / 4 2 - 6 1

    (b) / - * + 1 3 5 6 2

    (c) - 4 / + 1 * 3 2 7

    (d) * + 5 1 * - 4 2 / 9 3

6. Show how Listing 4.4 evaluates these postfix expressions, giving the stack contents after each step:

    (a) 1 6 4 5 * + 2 / -

    (b) 6 5 * 3 + 4 2 / -

    (c) 5 4 + 3 / 1 - 6 * 2 +

    (d) 6 2 1 + / 9 1 - 4 / -

7. Show how Listing 4.5 converts these expressions without parentheses to postfix, giving the stack contents after each step:

    (a) a + b + c + d

    (b) a + b * c - d + e

    (c) a * b + c * d - e * f

    (d) a / b / c + d * e * f

4.4 Evaluating Expressions: Implementations

The goal of this section is to implement the algorithms from Listings 4.4 and 4.5 as Java methods `evalPostfix()` and `toPostfix()`. The combination:

```
evalPostfix(toPostfix(expr))
```

will evaluate any infix expression. We assume all operands are integers, there are no parentheses in the infix expression (until the exercises), and that all terms are separated by spaces.

The `evalPostfix()` method is a little simpler than `toPostfix()`, and so it will be outlined in the exercises. For `toPostfix()`, we first isolate the task of determining an operator's precedence.

Operator Precedence

One way to determine operator precedence is to define a rank function so that

$$\text{rank}(t) = \begin{cases} 2 & \text{if } t \text{ is } * \text{ or } / \\ 1 & \text{if } t \text{ is } + \text{ or } - \\ -1 & \text{otherwise} \end{cases}$$

Then $\text{rank}(t) > \text{rank}(u)$ means that t has higher precedence than u, and $\text{rank}(t) = \text{rank}(u)$ if t and u have the same precedence. This rank function can also be used to determine whether a term is an operator or operand, since only operators have rank greater than zero. Zero is reserved for later use with parentheses (see Exercise 6).

As of Java 7, strings are allowed as **switch** expressions, and so Listing 4.6 uses that technique to implement the rank function.

Listing 4.6: Operator Rank

```java
1   public class Expression {
2       private static int rank(String op) {
3           switch (op) {
4               case "*":
5               case "/":
6                   return 2;
7               case "+":
8               case "-":
9                   return 1;
10              default:
11                  return -1;
12          }
13      }
14  }
```

toPostfix() Implementation

Listing 4.7 uses the **rank()** function to implement the infix-to-postfix conversion algorithm of Listing 4.5. It also uses the named constant **SPACE** to help improve readability. Its code should be added to the **Expression** class of Listing 4.6, in the same folder as your stack implementations.

Most of the code in the **toPostfix()** method should look familiar: it is based on the pseudocode in Listing 4.5 and has a structure much like the **acronym()** function in Listing 1.4, using a loop over tokens in a string and a **StringBuilder** to efficiently accumulate the result.

Listing 4.7: Infix to Postfix

```
1   // Add to Expression class
2   private static final String SPACE = " ";
3
4   public static String toPostfix(String expr) {
5       StringBuilder result = new StringBuilder();
6       Stack<String> operators = new ArrayStack<>();
7       for (String token : expr.split("\\s+")) {
8           if (rank(token) > 0) {
9               while (!operators.isEmpty() &&
10                      rank(operators.peek()) >= rank(token)) {
11                  result.append(operators.pop() + SPACE);
12              }
13              operators.push(token);
14          } else {
15              result.append(token + SPACE);
16          }
17      }
18      while (!operators.isEmpty()) {
19          result.append(operators.pop() + SPACE);
20      }
21      return result.toString();
22  }
```

Using Conditional Boolean Expressions

The difficult step in Listing 4.7 is the test in the while-loop on lines 9 and 10:

```
while (!operators.isEmpty() &&
        rank(operators.peek()) >= rank(token)) { ... }
```

The key to this expression is its use of the conditional && (AND), see page 4. The second half of the && is the main test:

```
rank(operators.peek()) >= rank(token)
```

If the operator at the top of the stack has the same or higher precedence, it needs to be popped, which happens in the body of the loop. The first part of the expression protects the peek() in case the stack is empty: remember that code using a stack is responsible for not peeking when the stack is empty.

Because the && is conditional, the second half is evaluated only if the first half is true, in which case the stack is not empty.

evalPostfix() Implementation

The algorithm to evaluate postfix is similar to the translation in `toPostfix()`, and you will be asked to implement it in the exercises. To do that, you will need to convert string operands to their integer value. Table 4.5 lists convenience methods from the wrapper class `Integer` to convert in both directions.

TABLE 4.5:   Integer Conversion Methods

static int parseInt(String s)
Converts string to decimal integer.
static String toString(**int** n, **int** b)
Converts integer n base b to a string.

Exercises

1. Explain why there are no break statements in the switch statement of Listing 4.6.

2. Modify Listing 4.6 to use named constants such as PLUS and MINUS for the operator symbols.

3. Write a `main()` method to test the `toPostfix()` method of Listing 4.7.

4. Add a static `isOperator()` method to the `Expression` class that returns true if the precedence of the operator is greater than 0. Rewrite the `toPostfix()` method to use `isOperator()`.

5. Modify the `Expression` class so that `toPostfix()` allows exponentiation using ^ in the infix expression. Give exponentiation a higher precedence than the other four arithmetic operations.

6. Modify the `Expression` class so that the `toPostfix()` method handles correctly balanced parentheses in the infix expression. Give parentheses rank 0, and modify the main for-loop to check for parentheses:

   ```
   if left paren, push onto stack
   if right paren
       pop and copy operators to output until left paren
       pop the left paren
   ```

   Do not copy any parentheses to the output, and put spaces around parentheses in infix expressions so they are seen as separate tokens.

7. Write an `applyOperator(operator, op1, op2)` method that takes an operator with two operands and returns the value of the binary computation. For example,

   ```
   applyOperator("*", 3, 7)
   ```

   should return 21. Use a switch statement, and assume all operands are integers.

8. Write the `evalPostfix()` method for the `Expression` class. Use the `applyOperator()` method from Exercise 7.

9. Modify Exercise 8 to perform exponentiation, using `Math.pow()` to do the calculation. You will need to cast the result back to an `int`.

10. Using Exercise 8, write an `eval()` method for the `Expression` class that evaluates an infix expression by first converting it to postfix and then evaluating the postfix. For example,

    ```
    eval("1 + 2 * 3")
    ```

    should return 7.

11. A stack can be used to determine whether or not an algebraic expression has properly balanced parentheses. For example,

Balanced	**Not Balanced**
$(x + y * (z - w))$	$(x + y * (z - w)$
$((x + (y)) * (z - w))$	$x + y * (z - w))$

    The idea is to scan each character in the expression, and if the character is a left parenthesis, push it on the stack. If it is a right parenthesis, then there should be a corresponding left parenthesis on the stack to pop. All other characters can be ignored.

    (a) Write the algorithm idea as pseudocode that returns true if the expression is balanced and false otherwise.

    (b) Implement the parenthesis matching algorithm as a method in the `Expression` class.

    (c) Extend the previous exercise to write a different method in the `Expression` class that checks parentheses (), square brackets [], and braces {}. Note that different types of parentheses need to be properly nested: for example, $(x + y * [z - w])$ is nested correctly, but $(x + y * [z - w)]$ is not.

Chapter 5

Queues

5.1   Interface and Linked Implementation

The close counterpart of a stack is a **queue**. In (primarily) British usage, a queue is a waiting line. In computer science, a queue is a data structure that acts like a waiting line, in which items are removed from the front of the line, and new items are added to the rear of the line. A queue is a **first-in, first-out (FIFO)** data structure, because the first element in will be the first one out.

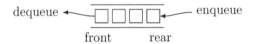

Adding to the rear of a queue is called **enqueuing**, and removing from the front of a queue is called **dequeuing** (pronounced "DQ-ing"). The front is drawn on the left here to match the linked lists we are about to consider.

Queue ADT

The Queue abstract data type in Table 5.1 is very similar to the Stack ADT. The exercises ask you to write the ADT as a Java interface.

TABLE 5.1:   Queue ADT

E dequeue()
Returns and removes item from front of queue.
void enqueue(E item)
Adds item to rear of queue.
boolean isEmpty()
Returns true if queue has no elements in it.
E peek()
Returns item at front of queue without removing it.
int size()
Number of elements in queue.

Like a stack, a queue is a **linear structure**, meaning that elements are stored in an ordered sequence, and so queues may be implemented with either arrays or linked lists. As with stacks, the goal is to make all operations $O(1)$ if we can. The linked implementation is somewhat simpler than the array version, so we begin with it.

Linked Implementation

Since a linked list has a natural "front" at the head, we begin by imagining the front of the queue at the beginning of the list:

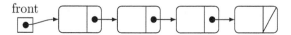

Then we need to work out how to efficiently provide the two main operations in the ADT:

Dequeuing will be easy in $O(1)$ time, because removing from the beginning of a linked list is just a single reference change. It is the same as popping from a linked stack.

Enqueuing is more difficult because we need to add new elements to the end of the list. Moving to the end of the list (known as **traversing** the list) will take $O(n)$ time, so that should be avoided. We can enqueue efficiently if we keep a separate **tail pointer** pointing to the last node in the list.

The last node in a queue is the rear of the list, so we keep a second instance variable named **rear**:

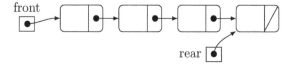

Enqueuing a new item will then require adding a new node after **rear**.

Insertion After a Node

Because it is no more difficult, we solve the more general problem of adding a new node after any given node **p**, not just after the last node.

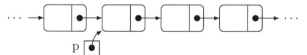

The steps are similar to those for inserting at the front of a linked list (see page 58):

1. Create a new node:

    ```
    Node q = new Node<>(item, null);
    ```

2. Set the **next** field of the new node to point to the node after p:

    ```
    q.next = p.next;
    ```

    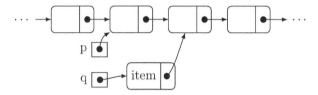

3. Change **p.next** to point to the new node:

    ```
    p.next = q;
    ```

    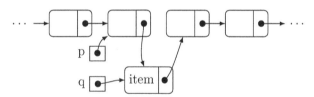

As before, these steps can be combined into one:

```
p.next = new Node<>(item, p.next);
```

Deletion After a Node

Although we will not need this operation for queues, deleting the node after a given node p is similar to inserting a new node after p because it is also just a change to **p.next**. Instead of pointing to a new node, though, **p.next** is updated to refer to the node that follows, which is pointed to by **p.next.next**.

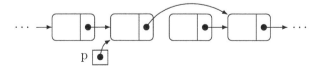

```
p.next = p.next.next;
```

The node to be deleted, pointed to by **p.next**, must exist, or the reference to **p.next.next** will cause a null pointer exception.

Managing the Tail

The only remaining issue is to be careful to update the rear pointer during enqueuing and dequeuing. The rear pointer begins null, like the front.

Enqueue If the queue is empty, create a node and set both **front** and **rear** to point to it. Otherwise, add a new node after **rear** and then update **rear** to point to the new node.

Dequeue Items are removed from the front, so the only time the rear pointer is involved is if the last item is being dequeued. In that case, **rear** should be set to null.

You will be asked to implement the **LinkedQueue<E>** class in the exercises.

Exercises

1. Show the results of these operations on an initially empty integer queue named q. Draw the queue contents after each operation, making clear where the front is, and give the return value of each non-void method.

   (a) q.enqueue(5)
       q.enqueue(8)
       q.peek()
       q.enqueue(3)
       q.dequeue()
       q.enqueue(10)
       q.size()
       q.enqueue(4)
       q.dequeue()
       q.dequeue()

   (b) q.enqueue(1)
       q.dequeue()
       q.enqueue(2)
       q.dequeue()
       q.enqueue(3)
       q.dequeue()
       q.size()
       q.enqueue(4)
       q.peek()
       q.dequeue()

   (c) q.enqueue(10)
       q.enqueue(20)
       q.enqueue(30)
       q.enqueue(40)
       q.peek()
       q.enqueue(50)
       q.dequeue()
       q.size()
       q.dequeue()
       q.dequeue()

2. Decide whether or not a queue would be an appropriate data structure for each of these types of task lists. Assume that when a new task arrives, it would be enqueued, and when a task is chosen to work on, it would be dequeued. Explain your answers.

   (a) Tasks that may need to be done in any order.

   (b) Tasks where the next one to work on is always the one that has been waiting the longest.

   (c) Tasks where the next one to work on is always the most recently received.

   (d) Tasks that need to be done in the order they are received.

   (e) Tasks that may need to be shuffled or sorted.

3. Write a generic Java interface `Queue<E>` for the Queue ADT in Table 5.1.

4. Develop the `LinkedQueue<E>` class, implementing the `Queue<E>` interface from Exercise 3 with a linked list as described in this section. Write these members:

   (a) Private `Node<T>` class Include a second constructor that only takes one parameter, the data item, and sets the `next` field to null.

   (b) `dequeue()` Throw a `NoSuchElement` exception from `java.util` if the queue is empty (see page 55).

   (c) `enqueue()`

   (d) `isEmpty()`

   (e) `peek()` Throw a `NoSuchElement` exception if the queue is empty.

   (f) `size()` Include a `size` instance variable and update it where necessary.

5. Add a `main()` method to the `LinkedQueue<E>` class to test the queue with:

   (a) Integers

   (b) Doubles

   (c) Strings

6. Determine the $O()$ time complexity of these `LinkedQueue` operations. Explain your answers.

   (a) `dequeue()`   (d) `peek()`

   (b) `enqueue()`   (e) `size()`

   (c) `isEmpty()`

7. As described above, the enqueue method has to treat an empty queue as a special case. To avoid this, a **dummy node** can be created at the front of the list containing no data. This simplifies enqueuing, but requires rewriting other methods to take the dummy node into account. Write a `DummyLinkedQueue<E>` implementation of `Queue<E>` using this approach. Discuss the tradeoffs.

5.2   Array Implementation

The linked queue was a bit more complicated than its stack counterpart, requiring a second tail pointer to enqueue efficiently. In a similar way, the array implementation of a queue will require extra work to achieve $O(1)$ performance. The main ideas are similar to an array stack, though, so we begin by ignoring the complications.

Basic Idea

As with the array implementation of a stack, it is helpful to begin by imagining a few elements being added to a queue. Suppose 10, 20, and 30 are enqueued in that order; then it is natural to store them at the beginning of the array (shown as primitives to simplify the pictures):

0	1	2	3	4
10	20	30		

At this point, 10 is at the front of the queue and 30 is at the rear, so instance variables should track those locations:

0	1	2	3	4
10	20	30		

front: 0   rear: 2

If the next operation is to enqueue 40, then it goes to the rear of the queue:

0	1	2	3	4
10	20	30	40	

front: 0   rear: 3

Code to enqueue a new **item** like this is straightforward:

```
data[++rear] = item;
```

If the next operation is dequeue, then 10 should be returned and then `front` increased:

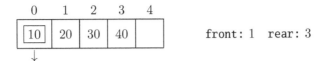

front: 1   rear: 3

and so the code to dequeue is:

```
return data[front++];
```

Problem: Drift

The problem with this basic idea is that the queue gradually moves to the right in the array. Even if the queue never has more than a few elements in it, if there is a lot of enqueuing and dequeueing, the queue will eventually fall off the end of the array:

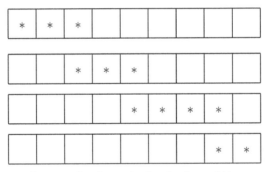

Shifting the queue elements back to the beginning of the array would take a lot of time if the array was large, and increasing the size of the array would be wasteful for only a handful of items. A better solution is to teach the array to wrap around.

Circular Array

A **circular array** wraps around, so that incrementing the last index wraps back around to the first index:

In the above example, that would mean "incrementing" 6 to get 0 instead of 7.

The remainder operator % does this nicely:

```
(6 + 1) % 7 = 7 % 7 = 0
```

and so the general code to increment in a circular array is:

```
var = (var + 1) % array.length
```

This will be a common operation in the queue, so it might be worth keeping a `capacity` instance variable storing the current length of the array. Then updating `front` and `rear` become:

```
rear = (rear + 1) % capacity;
front = (front + 1) % capacity;
```

Modifying the basic operations:

```
data[++rear] = item;        // enqueue
return data[front++];       // dequeue
```

to use circular increments instead of `++` will solve the drifting problem.

Resizing Circular Arrays

By solving one problem we have created another: resizing now has to take circularity into account. To see the problem, suppose a queue has been in use for some time, has elements 10, 20, ..., 50 queued in that order, and is now full:

0	1	2	3	4
40	50	10	20	30

front: 2   rear: 1

If we resize in the same way as an array stack (see Listing 3.2), then the result would be:

0	1	2	3	4	5	6	7	8	9
40	50	10	20	30					

front: 2   rear: 1

This will not work: the queue expects 40 to follow 30 (because it had in the circular array), and even though there are empty slots, there is no room after the end of the queue (i.e., after item 50). Therefore, we need to change the resizing method so that the front of the queue is back in slot 0:

0	1	2	3	4
40	50	10	20	30

front: 2   rear: 1

0	1	2	3	4	5	6	7	8	9
10	20	30	40	50					

front: 0   rear: 4

This looks like it would be tricky to write, but it's not too bad with the right organization. The idea is to loop over the new array locations with one variable i, while updating a second variable j that loops through the old circular array. Listing 5.1 outlines this approach. Notice in it how the for-loop updates i, while j is updated for the old circular array inside the loop.

Listing 5.1: Resize Circular Array (Pseudocode)

```
1  j = front
2  for i = 0 to size - 1
3    newData[i] = data[j]
4    j = (j + 1) % capacity
5  Update front, rear, capacity, data
```

The exercises ask you to complete this implementation and the rest of the ArrayQueue<E> class.

Exercises

1. Explain the use of prefix and postfix increment in the basic array queue operations:

   ```
   data[++rear] = item;      // enqueue
   return data[front++];     // dequeue
   ```

2. Determine the correct starting values for the front and rear array indices of an array queue. Explain your answers.

3. In a circular array implementation of a queue:

   (a) Explain why this relationship is true for both empty and full queues:

       ```
       (rear + 1) % capacity == front
       ```

   (b) If there were no size instance variable storing the number of items in the queue, how might you tell the difference between an empty queue and a full queue? Outline a strategy.

4. Circular arrays were introduced in part to avoid shifting elements to reposition the queue at the front of the array. However, that is precisely what the resize() method does, because of circularity. Is that a problem? In other words, does resizing with Listing 5.1 impact the queue's performance differently than resizing an array stack? Explain why or why not.

5. Develop the `ArrayQueue<E>` class, implementing the `Queue<E>` interface with a circular array as described in this section. Use separate `size` and `capacity` instance variables, and write these methods:

   (a) `dequeue()`   Throw a `NoSuchElement` exception if the queue is empty, and set the obsolete reference (see page 71) to null.

   (b) `enqueue()`   Double the size of the array if the queue is full.

   (c) `isEmpty()`

   (d) `peek()`   Throw a `NoSuchElement` exception if the queue is empty.

   (e) `size()`   Include a `size` instance variable and update it where necessary.

   (f) Private `resize()`   Implement Listing 5.1.

   (g) Two constructors: one default, and one that takes an initial capacity as a parameter. The default constructor should call the other constructor using **this**().

6. Add a `main()` method to the `ArrayQueue<E>` class to test the queue with:

   (a) Integers

   (b) Doubles

   (c) Strings

7. Determine the $O()$ time complexity of these `ArrayQueue` operations. Explain your answers.

   (a) `dequeue()`

   (b) `enqueue()`   Ignore resizing.

   (c) `isEmpty()`

   (d) `peek()`

   (e) `size()`

5.3   Inheritance: Fixed-Length Queues

Queues turn up in interesting places and are not always implemented in software. For example, **hardware queues** are common components in systems that receive data at a variable rate: the queue allows items to collect in a waiting line until the processor is ready for more input. Both graphics processing units and network routers use hardware queues to manage incoming data in this way.

Examples: GPUs and Network Routers

Graphics processing units (GPUs) are constructed with a pipeline architecture, which you can imagine as a sequence of stages connected by pipelines:

Graphics data move down these pipelines from one stage to the next, with hardware FIFO queues at each stage to prevent stalling when one of the stages cannot keep up with the inputs coming into it.

Hardware queues are also used in network routers to hold incoming packets before they are processed. Because of the importance of not dropping network packets, routers usually also have other mechanisms including software queues to handle full hardware queues.

In this section, we develop a **fixed-length queue** to model hardware queues or other queues that have a fixed maximum capacity. Given such a queue, attempting to enqueue into a full queue will result in the item being *dropped*, like a dropped network packet.

Inheritance

Rather than writing a fixed-length queue class from scratch, it would be better to take advantage of the work we've already done. The basic idea of **inheritance** is for one class (the **subclass**) to begin with the state and behavior of another class (the **superclass**), but then be allowed to add new variables and methods, as well as redefine how old methods work. The new class is said to **extend** the old class and so is also known as an **extension** of the **base class**.

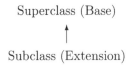

Inheritance Hierarchies

Sub- and superclass relationships between classes create an **inheritance hierarchy** with more general classes towards the top and more specific classes towards the bottom.

For example, some of the classes you have seen so far are related by inheritance in this way:

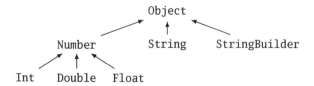

This means that the `Number` class, which we have not seen yet, captures common state and behavior of all of the numeric wrapper classes, as well as other numeric classes.

In fact, *every* Java class inherits from the `Object` class, either directly or indirectly. In other words, the `Object` class is at the root of the entire inheritance hierarcy. We will return to the role of the `Object` class at the end of this section.

Is-A and Has-A Relationships

Inheritance models an **is-a** relationship:

A `Double` is a `Number`

A `Number` is an `Object`

A `FixedLengthQueue` is a `Queue`

Another common relationship is **has-a**, which is used when an object has a component of that type:

An `ArrayQueue` has a `size`

A `Fraction` has a `numerator`

An array has a `length`

Has-a relationships are modeled by **composition**, which is simply using an instance variable to store the component. Deciding between these two types of relationships is an important part of software design.

Implementation

A fixed-length queue is close to an `ArrayQueue`, except that the size of its underlying array should never change. Thus, we implement it in Listing 5.2 as an extension of the `ArrayQueue` class. If an attempt is made to enqueue when the queue is full, instead of resizing the array, the item will be dropped and the queue will track the total number of dropped items.

Listing 5.2: Fixed-Length Queue

```
1   public class FixedLengthQueue<E> extends ArrayQueue<E> {
2       private int drops;
3
4       public FixedLengthQueue(int capacity) {
5           super(capacity);
6       }
7
8       @Override
9       public void enqueue(E item) {
10          if (size == capacity) {
11              drops++;
12          } else {
13              super.enqueue(item);
14          }
15      }
16
17      public int drops() {
18          return drops;
19      }
20  }
```

Java Class Extensions

The **extends** clause is used to declare a new class as a subclass of an existing class:

```
public class SubClass extends SuperClass { ... }
```

The subclass then inherits the class members (fields, methods, and nested classes) from the specified superclass. New instance variables and methods may also be declared inside the subclass, as on line 2 and beginning at line 17 of Listing 5.2.

If the superclass implements an interface, then so does the subclass since it inherits the public members of the superclass.

Subclass Constructors

If you write a constructor for a subclass, the first thing it must do is call a constructor from the superclass using **super**:

```
super(arguments);
```

The only time this is not required is if the constructor is calling a different constructor of the subclass using **this**(). The **super**() call in the constructor on line 5 assumes that a constructor with an integer capacity parameter is part of the ArrayQueue class (see Exercise 4).

Overriding Methods

Redefining a method from the superclass in a subclass is called **overriding** the method. In this case, the enqueue() method is overridden beginning at line 9 in order to check the size before actually enqueuing the item. The @Override annotation signals to the compiler and readers that we intend to override an existing method rather than define a new one.

Existing superclass methods may be called inside the body of an overridden method using **super**:

```
super.method(arguments)
```

In Listing 5.2, once the enqueue() method has determined that the queue will not overflow, the superclass method is called on line 13.

Protected Fields in the Superclass

One other change (in addition to making sure there is a constructor with a capacity parameter) needs to be made to the existing ArrayQueue superclass in order for this subclass to compile. As it is, the subclass cannot use the instance variables size and capacity because they were (or should have been) declared private in ArrayQueue. Class members declared private cannot be used *anywhere* outside of their class, even in a subclass.

Because it is reasonable for a subclass to access some instance variables or internal methods, Java provides a third visibility option, **protected**, for this situation. Protected class members may be accessed by subclasses as well as the class itself, but not by outside classes.[1]

The Object Class

Table 5.2 lists two of the most important methods defined in the Object class. Because every class is an extension of the Object class, every Java object—including arrays—inherits these methods, along with others listed in the documentation.

---

[1]Protected members are also accessible by classes in the same package, but named packages are not used in this text.

TABLE 5.2: Object Methods

boolean equals(Object o)
True if this object and o are the same object.
String toString()
Returns string representing this object.

The default meaning of equals() for Objects is the same as ==, but classes may override equals() to define their own notion of equality.[2] The signature of an overridden method must not change, so this explains why the String equals() method (see Section 1.2) takes an Object parameter instead of a String.

The Object implementation of toString() returns the class name and the object's hashcode (see Section 11.2) in hexadecimal. When a class like the Fraction class in Section 1.5 defines a toString() method, it is overriding the Object definition.

Finally, recall from Section 4.2 that the Object class was used to create generic arrays, which were then cast to the type E[]. The reason this works is that no matter what E is, it inherits from Object, and so an E object "is a" Object.

Exercises

1. Use Listing 5.2 to:

    (a) List the new instance variables (if any) in the FixedLengthQueue class.

    (b) List the new methods (if any) in the FixedLengthQueue class.

    (c) List the methods (if any) from ArrayQueue that have been overridden.

2. Describe the difference between overloading and overriding a method in Java.

3. Explain why we chose to have FixedLengthQueue extend ArrayQueue rather than LinkedQueue. Could extending LinkedQueue be made to work? Explain why or why not.

---

[2]Overriding equals() is not as straightforward as it might seem. See Items 8 and 9 from Bloch [4] for details.

4. Modify `ArrayQueue` to work with the `FixedLengthQueue` class by making these changes:

   (a) Implement Exercise 5g from Section 5.2 if you have not done so already to have an `ArrayQueue` constructor with an integer capacity parameter.

   (b) Change the visibility of `size` and `capacity` so they may be used by the subclass.

5. Decide whether inheritance, composition, neither, or both seem appropriate for each of these related classes. If inheritance fits, indicate the subclass and superclass; if composition, indicate which object is the instance variable in the other class. Explain your answers.

   (a) A `Person` class and an `Employee` class in a human resources system.

   (b) An `Employee` class and a `Department` class in a human resources system.

   (c) An `Employee` class and a `Manager` class in a human resources system.

   (d) A `Shape` class and a `Sphere` class in a graphics system.

   (e) A `Cube` class and a `Sphere` class in a graphics system.

6. Add a `main()` method to the `FixedLengthQueue<E>` class to test the queue with:

   (a) Integers

   (b) Doubles

   (c) Strings

7. Determine the $O()$ time complexity of these `FixedLengthQueue` operations. Explain your answers.

   (a) `dequeue()`

   (b) `enqueue()`

   (c) `isEmpty()`

   (d) `peek()`

   (e) `size()`

8. Consider the following code. Does it compile and run? If so, give the output; if not, explain why not and how to fix it.

```
Queue<Integer> q = new FixedLengthQueue<>(50);
for (int i = 0; i < 100; i++) {
    q.enqueue(i);
}
System.out.println(q.drops());
```

9. Using inheritance, write a `FixedLengthStack<E>` class to represent a fixed-size stack that drops pushed elements if the stack is full. Include any necessary changes to the superclass.

Project: Fixed-Length Queue Simulation

A natural question arises with fixed-length queues: how important is the size of the queue under heavy traffic? If many items are being dropped, is it worth investing in a larger queue? If so, how large? This project develops a **simulation** to help answer questions like these.

Model

Imagine a generic "processor" with an input queue that receives tasks and processes them, like the stages in a GPU or a network router. Each task has a given integer length of time that it will take to complete on the processor. If the processor is busy with one task and receives another, it uses a fixed-length queue to store the waiting tasks. Once a processor finishes a task, it begins working on the next task in its queue. A processor with no work to do is considered free.

Table 5.3 lists a set of public methods for a simulated processor. Tasks are represented simply by the integer amount of time they take to complete.

TABLE 5.3:   Simulated Processor

void `addTask(`**int** `taskTime)`
Accepts new task of given length of time. If free, this becomes the current task; otherwise, adds task to queue.
int `drops()`
Number of items dropped by queue.
boolean `free()`
True if not processing a task and queue is empty.
void `tick()`
Simulates one time step. If busy, does one unit of work on current task. If current task finishes, checks queue for next one to start.

Clock-Based Simulations

Simulations often use an integer counter to simulate a clock, where incrementing or decrementing the counter corresponds to one tick of the clock. This processor simulation will randomly generate new tasks for a certain amount of time minTime and then continue running until all tasks are finished. Listing 5.3 describes one way to organize the simulation with a clock variable t.

Listing 5.3: Clock-Based Simulation (Pseudocode)

```
1  t = 0
2  while (t < minTime or processor is busy)
3      if (t < minTime) randomly add new task to processor
4      simulate one time step on processor
5      t++
```

Random Tasks

Simulation-based programs use a variety of techniques to generate random sequences of events. We will use the following simplified model to simulate a random series of tasks for the processor: during each time step of the simulation, there is probability taskChance (between 0 and 1) of creating a new random task. Each task will require processing time of some random integer number of steps between 1 and a fixed upper limit maxTaskTime. Listing 5.4 outlines this procedure in pseudocode.

Listing 5.4: Generate Random Tasks (Pseudocode)

```
1  r = random double from [0, 1)
2  if (r < taskChance)
3      taskTime = random int from 1 to maxTaskTime
4      add task with length taskTime to processor
```

Exercises

1. Write a Processor class to represent a single processor, implementing the public methods from Table 5.3. The Processor constructor should take a parameter specifying the capacity of its fixed-length queue.

2. Write a `QueueSimulator` class to drive the simulation of one processor. Include one public method:

> **void** run(**int** queueLength, **int** minTime, **int** maxTaskTime,
>   **double** taskChance)

that implements Listing 5.3. It should call a separate method implementing Listing 5.4.

At the end of the `run()` method, display the number of tasks processed, the percentage of tasks dropped, and the amount of extra time (beyond `minTime`) needed for the processor to finish. Consider the `System.out.format()` method for nice output.

3. Use the `QueueSimulator` from the previous exercise to explore a situation when overcrowding is predictable, such as with `taskChance = 0.3` and `maxTaskTime = 10`. Use `minTime = 10000` and queue lengths between 5 and 500. Discuss the impact of different queue lengths according to the simulator. Is it what you expected?

Chapter 6

Lists

The power of stacks and queues as linear data structures lies in their limited interface: elements may only be added or removed in specific ways. Controlling insertion and deletion in this way allows designing for $O(1)$ performance for those operations.

Lists, on the other hand, provide a very flexible interface that allows inserting and removing elements anywhere in the list. This flexibility comes at a price, though: operations will no longer necessarily be $O(1)$.

6.1 Interface

The List ADT views its data much like an array does: elements are accessible via consecutive indices.

List Indexing

Think of references in a list as stored at indices 0 through one less than the number of elements in the list, just like an array:

This description implies:

Lists are dynamic because they continually grow and shrink as items are added and removed. The length of a list always depends on the number of elements in it and is not fixed like an array.

There are never gaps between items in a list because it is defined to use indices 0 through $\texttt{size} - 1$.

Be aware that the list may not be implemented with an array, so this diagram is just a way to think about how the elements are organized by the ADT. In fact, a linked implementation will be developed in Section 6.3.

List ADT

Whereas the stack or queue ADTs are fairly standard, the List abstract data type has many variations and can be quite complex. We will use the set of methods in Table 6.1.

TABLE 6.1:   List ADT

void add(E item) Adds item to end of list.
void add(**int** index, E item) Inserts item at index, shifting items to right to make room. If index = size, adds item to the end of the list.
E get(**int** index) Returns item at index.
int indexOf(E item) Index of first occurrence of item in list, −1 if not found.
boolean isEmpty() True if list has no elements.
E removeAt(**int** index) Removes and returns item at index, shifting remaining items to left.
E set(**int** index, E item) Replaces item at index, returning the previous value.
int size() Number of elements in list.

Shifting

Stacks and queues limit access to their contents, either through the top (stack) or front and rear (queue). Because the List ADT allows accessing any element at any time, some list methods affect the indices of other items in the list. Both add() and removeAt() cause elements other than those being added or removed to change position. For example, if a list contains

    5, 8, 2, 1, 4, 7

then the operation add(3, 6) results in the list

    5, 8, 2, 6, 1, 4, 7

The elements after 6 all now have a different index than they did prior to the add(). Similarly, removeAt(1) returns 8 (the item in slot 1) and modifies the list to be

    5, 2, 6, 1, 4, 7

Again, remember that the list may not be implemented with an array, so no actual shifting may occur. What changes from the perspective of the ADT are the index locations of some of the elements in the list.

Overloading Methods

The List ADT overloads the `add()` method since two versions are given. Overloading is useful when more than one set of parameters may be convenient for what is essentially the same method. In this case, adding an item to the end of a list will likely be common, so it is given a simplified method call.

Arrays and Lists

In a sense, a list is a more powerful array that will grow as needed and shift elements for insertion and deletion. And because it's an ADT, a list can be implemented in a variety of ways to provide different performance characteristics.

In fact, most Java programming is done with lists from the `java.util` library rather than arrays. The reason we have been using arrays is to study the low-level implementation details of different data structures. Once you know how those structures work, implementations of the List ADT are more powerful and convenient.

Exercises

1. Show the results of these operations on an initially empty integer list named `items`. Indicate the return value of all non-void methods and draw the list contents after each operation.

(a) `items.add(5)`
    `items.add(8)`
    `items.add(3)`
    `items.removeAt(1)`
    `items.add(10)`
    `items.add(4)`
    `items.removeAt(3)`
    `items.removeAt(1)`

(b) `items.add(1)`
    `items.set(0, 2)`
    `items.add(3)`
    `items.set(0, 4)`
    `items.set(1, 5)`
    `items.add(6)`
    `items.removeAt(1)`
    `items.get(1)`

(c) `items.add(10)`
    `items.add(0, 20)`
    `items.add(1, 30)`
    `items.add(0, 40)`
    `items.add(2, 50)`
    `items.indexOf(10)`
    `items.indexOf(20)`
    `items.indexOf(30)`

2. Suppose `items` is a List containing {1, 7, 2, 4, 1, 8, 9, 7}. Show the result of these operations, run in this sequence, and give the final contents of the list.

   ```
   items.size()
   items.get(3)
   items.indexOf(7)
   items.removeAt(5)
   items.add(0, 3)
   items.add(2, 5)
   items.removeAt(1)
   ```

3. Decide whether or not a list would be an appropriate data structure for each of these types of task lists. Explain your answers.

   (a) Tasks that may need to be done in any order.

   (b) Tasks where the next one to work on is always the one that has been waiting the longest.

   (c) Tasks where the next one to work on is always the most recently received.

   (d) Tasks that need to be done in the order they are received.

   (e) Tasks that may need to be shuffled or sorted.

4. Write a generic Java interface `List<E>` for the List ADT.

5. Suppose `ArrayList<E>` implements the `List<E>` interface from Exercise 4.

   (a) Write Java code to declare and create an integer array list `items`, and then add the values 0 through 9 to `items` (in that order) so that they are stored in increasing order.

   (b) Write Java code to declare and create an integer array list `items`, and then add the values 0 through 9 to `items` (in that order) so that they are stored in decreasing order.

6. Describe how to implement push and pop operations for a stack using the List ADT, trying to avoid shifting elements if at all possible. If it is not possible to avoid shifting, explain why.

7. Describe how to implement enqueue and dequeue operations for a queue using the List ADT, trying to avoid shifting elements if at all possible. If it is not possible to avoid shifting, explain why.

6.2 Array Implementation

With its emphasis on indexing, an array implementation of the List ADT should seem natural. The `add()` method developed below illustrates how to address two issues that did not arise with stacks and queues: checking indices and shifting to make room for the new item. Exercises ask you to finish the `ArrayList` implementation.

Index Checking

By design, elements in a list have indices 0 through the one less than the current size. Therefore, every method in the List ADT with an index parameter has the responsibility to check that the index is valid before performing its operation. Any invalid index should result in an `IndexOutOfBoundsException`.

The `add()` method is slightly different than other methods that only access existing elements, since it is allowed to use the size of the list as an index.

Shifting in an ArrayList

Two types of shifting are needed in an array list: making room for a new item in `add()`, and closing the gap created by `removeAt()`. To get the code right, pay attention to where the open slot is and decide which element has to move first.

For `add(index, item)`, assuming there is room for the new item, the opening is on the right, and so the item in slot `size - 1` has to move first:

This suggests starting the loop at `size - 1` and decrementing until `index`:

```
for (int i = size - 1; i >= index; i--) {
    data[i + 1] = data[i];
}
```

In that case, the index i represents the index being copied from. An alternative is to loop from `size` down to `index + 1`, which would give the index being copied to.

To close the gap in `removeAt(index)`, the opposite needs to happen:

In this case, the opening is on the left, and the item at `index + 1` needs to move first. The details for this case are left as an exercise.

Listing 6.1 demonstrates shifting and index checking for the `ArrayList add()` method.

Listing 6.1: ArrayList Insertion

```java
public void add(int index, E item) {
    if (index < 0 || index > size)
        throw new IndexOutOfBoundsException();
    if (size == data.length) resize(2 * data.length);
    // shift right to make space
    for (int i = size - 1; i >= index; i--) {
        data[i + 1] = data[i];
    }
    data[index] = item;
    size++;
}
```

Exercises

1. Explain how the array list `add()` method in Listing 6.1 works when `index` equals `size`; i.e., when adding to the end of the list.

2. Develop the `ArrayList<E>` class implementing the `List<E>` interface by writing these methods:

   (a) `add(item)`   Double the length of the array if it is full.

   (b) `get()`

   (c) `indexOf()`   Use linear search to find the first occurrence of the item.

   (d) `isEmpty()`

   (e) `removeAt()`   Set the obsolete reference (see page 71) to null.

   (f) `set()`

3. Exercise 2a can be written in two ways, either from scratch or by calling the `add()` method in Listing 6.1. Write both versions and discuss the differences.

4. Modify the `removeAt()` method from Exercise 2e to reduce the array length by half if the number of elements in the list is less than or equal to one-fourth the current length of the array, but do not let the array have a length smaller than 10.

5. Override the `toString()` method in the `ArrayList` class by calling the `toString()` method of each list element and accumulating the results in a `StringBuilder`.

6. In the previous exercise, explain how you know `toString()` can be called on every list item if the list is generic.

7. Add a `main()` method to the `ArrayList` class to test the list with:

   (a) Integers
   (b) Doubles
   (c) Strings

8. Determine the $O()$ performance of each of these `ArrayList` methods. Explain your answers.

   (a) `add(item)`   Ignore resizing.
   (b) `add(index, item)`   Ignore resizing.
   (c) `get()`
   (d) `indexOf()`
   (e) `isEmpty()`
   (f) `removeAt()`   Ignore resizing from Exercise 4.
   (g) `set()`
   (h) `size()`

9. Assuming `items` is an `ArrayList`, determine the $O()$ performance of each of these fragments of code. Give your answers in terms of n, the number of elements in `items`.

   (a)
   ```java
   for (int i = 0; i < items.size(); i++) {
       System.out.println(items.get(i));
   }
   ```

   (b)
   ```java
   while (!items.isEmpty()) {
       System.out.println(items.removeAt(0));
   }
   ```

   (c)
   ```java
   while (!items.isEmpty()) {
       System.out.println(items.removeAt(items.size() - 1));
   }
   ```

10. Assuming `items` is an empty `ArrayList`, determine the $O()$ performance of each of these fragments of code. Give your answers in terms of **n**.

(a)
```
for (int i = 0; i < n; i++) {
    items.add(i);
}
```

(b)
```
for (int i = 0; i < n; i++) {
    items.add(0, i);
}
```

6.3   Linked Implementation

In contrast to the array implementation of the List ADT, a linked list implementation may seem less natural. Linked lists are good at moving from one node to the next rather than directly to a particular index. However, keep in mind that the array implementation has its own awkwardness: any time an element is inserted or removed from the list, other items have to shift in the array. A linked implementation never has to shift elements, since it can simply reassign pointers.

Two linked list options are helpful for the List ADT: using a dummy node and double links.

Dummy Nodes

Compare the steps for inserting and removing at the front of a linked list (pages 58–60) with inserting and removing after a node (pages 86–88). The front of the list is different from everywhere else because it requires changing the head of the list.

Unlike stacks and queues, the List ADT allows inserting and removing anywhere in the list, so it would be helpful to not have to write a special case for the front of the list. We can do this by creating a **dummy node** at the front of the list:

Then every insertion or deletion will occur "after a node," and the head will never change. The dummy node contains no data; its purpose is just to simplify insertion and deletion code. If a dummy node is used, other nodes are sometimes referred to as **regular nodes**.

Double Links

The linked lists we have used to this point are known as **singly-linked** lists because each node has a single link to the next node. A **doubly-linked** list has two pointers in each node, one to the previous node and one to the next:

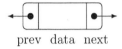

prev data next

Single links were sufficient for stacks and queues (and are also sufficient for the List ADT), but there is some convenience in using double links for insertion and deletion in the middle of a list.

Be careful, though: double links make it easier to write code that generates null pointer exceptions. The reason is that compound expressions such as `p.next.prev` and `p.prev.next` become common. The key to writing such expressions was given in Section 3.3: make sure that variables being dereferenced are not null. In a compound expression, there are two references to check:

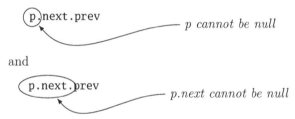

p cannot be null

and

p.next cannot be null

Deletion with Double Links

To illustrate the use of double links, we develop the code to delete a given node p.

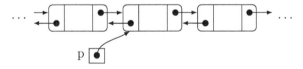

To avoid compound expressions for now, assign references to the node before p and the node after p:

```
Node<E> prev = p.prev;
Node<E> next = p.next;
```

The main step is the same as for deleting the node after `prev` (see page 87):

```
prev.next = next;
```

Updating `prev.next` results in this change to the diagram:

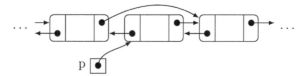

If there is a dummy node, that step is safe, because `prev` must exist. However, there is no guarantee that `next` is not null, and so the second connection requires a check first:

```
if (next != null) next.prev = prev;
```

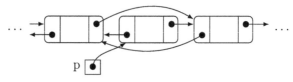

The complete `removeAt()` method is in Listing 6.2.

Traversals

Walking through the nodes of a linked list is known as **traversing** the list. The idea of a full traversal is simple: start at the front of the list (or perhaps skip the dummy node) and keep following `next` pointers until you reach null. The Java implementation is a nice for-loop:

```
for (Node<T> p = head; p != null; p = p.next) {
    // work with node pointed to by p
}
```

Recall that the three components of the **for** are initialize, test, and increment, and those are exactly the steps needed for traversal.

Indexed Access

Because linked lists do not have direct indexed access to each element, it will be helpful to write a private `getNode()` method that returns the node for a given index. The `getNode()` method does a partial traversal until reaching the desired index.

Listing 6.2 gives the implementation of `getNode()` and illustrates its use in `removeAt()`. Rather than skipping the dummy node, the index -1 is allowed in case the dummy node is needed by other methods. Because `getNode()` will only be called by code internal to the class, it can assume the target index is valid. Remember that nodes are an implementation detail of the linked list and so should not be returned by public methods.

Listing 6.2: Linked List Deletion and Helper

```
1   public E removeAt(int index) {
2     if (index < 0 || index >= size)
3           throw new IndexOutOfBoundsException();
4     Node<E> p = getNode(index);
5     Node<E> prev = p.prev;
6     Node<E> next = p.next;
7     E result = p.data;
8     prev.next = next;
9     if (next != null) next.prev = prev;
10    size--;
11    return result;
12  }
13
14  private Node<E> getNode(int index) {
15    // assumes -1 <= index < size.   -1 is dummy.
16    Node<E> p = head;
17    for (int i = -1; i < index; i++) {
18        p = p.next;
19    }
20    return p;
21  }
```

Performance

As an example of analyzing linked list operations, consider the performance of the removeAt() method in Listing 6.2 in terms of n, the number of items in the list. It appears to be a simple sequence of a fixed number of steps, which would be $O(1)$. But one of those steps is a call to getNode(), which traverses the list. Thus, the performance of removeAt() is $O(n)$ because the traversal may take up to n steps.

Using Nodes and References

One last piece of advice for navigating linked list code: at times, you will need to distinguish between a reference and the node it refers to. For example, using single links for simplicity, we often view the node after the node referenced by p "as" p.next:

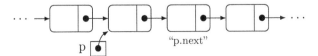

But if you need to change **p.next**, the reference **p.next** lives in the node pointed to by **p**, along with **p.data**:

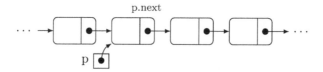

Understanding this difference between a reference and the node it refers to will help you write correct code.

Exercises

1. Explain why **getNode()** was not used to find **prev** and **next** in the **removeAt()** method of Listing 6.2. Could it have been used?

2. Describe the circumstances (i.e., which node or nodes are being deleted) when the if-statement at line 9 prevents a null pointer exception in Listing 6.2.

3. Rewrite the **removeAt()** method of Listing 6.2 to use compound expressions instead of the separate variables **prev** and **next**.

4. Develop the **LinkedList<E>** class implementing the **List<E>** interface by writing these class members. Use double links and a dummy node.

   (a) Private **Node<T>** class   Include a default constructor taking no parameters that sets all three fields to null.

   (b) **add(item)**

   (c) **add(index, item)**

   (d) **get()**

   (e) **indexOf()**. Use linear search to find the first occurrence of the item.

   (f) **isEmpty()**

   (g) **set()**

   (h) **size()**

5. Exercise 4b can be written in two ways, either from scratch or by calling the other **add()** method. Write both versions and compare them.

6. Override the **toString()** method in the **LinkedList** class by writing a traversal that calls the **toString()** method of each list element and accumulates the results in a **StringBuilder**.

7. Add a main() method to the LinkedList class to test the list with:

   (a) Integers

   (b) Doubles

   (c) Strings

8. Determine the $O()$ performance of each of these LinkedList methods. Explain your answers.

   (a) add(item)

   (b) add(index, item)

   (c) get()

   (d) indexOf()

   (e) isEmpty()

   (f) set()

   (g) size()

9. Use the previous exercise to compare the performance of LinkedList with ArrayList. Does one or the other have a clear advantage? Explain your reasoning.

10. Assuming items is a LinkedList, determine the $O()$ performance of each of these fragments of code. Give your answers in terms of n, the number of elements in items.

   (a)
```java
for (int i = 0; i < items.size(); i++) {
    System.out.println(items.get(i));
}
```

   (b)
```java
while (!items.isEmpty()) {
    System.out.println(items.removeAt(0));
}
```

   (c)
```java
while (!items.isEmpty()) {
    System.out.println(items.removeAt(items.size() - 1));
}
```

11. Assuming `items` is an empty `LinkedList`, determine the $O()$ performance of each of these fragments of code. Give your answers in terms of n.

    (a)
```
for (int i = 0; i < n; i++) {
    items.add(i);
}
```

    (b)
```
for (int i = 0; i < n; i++) {
    items.add(0, i);
}
```

6.4   Iterators

Exercise 10a in Section 6.3 exposes an unexpected flaw of the `LinkedList` implementation: even though traversing a linked list is efficient, it becomes inefficient if we use `get()` in a for-loop. In other words, the public interface of `LinkedList` provides no way to efficiently traverse the list. Iterators allow us to rectify this problem and support Java's enhanced for-loop at the same time.

Iterators

A Java iterator is an object designed for iterating through the items of a collection. The `java.util.Iterator` interface in Table 6.2 defines the functionality of an iterator.

TABLE 6.2:   java.util.Iterator Interface

boolean hasNext()   True if iterator has another element.
E next()   Returns next element and advances iterator.
void remove()   (Optional) Deletes last element returned by next(). May only be called once per call to next().

Code using an iterator should call `hasNext()` before calling `next()` to be sure there is an element available. If `next()` is called when `hasNext()` is false, a `java.util.NoSuchElementException` is thrown. Iterators that do not support the `remove()` method must have the `remove()` method throw an `UnsupportedOperationException`.

Iterables

A collection that can provide an iterator is called **iterable** and implements the
`Iterable` interface in Table 6.3. That means it provides a public `iterator()`
method returning an object that implements the `Iterator<E>` interface. This
object is often defined by a private inner class.

TABLE 6.3:   Iterable Interface

`Iterator<E> iterator()`
Returns an iterator for this collection.

The terminology of iterables can be difficult to follow, especially at first. De-
veloping a specific iterator for linked lists will help.

Linked List Iterator

The problem with repeatedly calling `get()` to traverse a linked list is that
`getNode()` has to start back over at the beginning of the list every time. The
linked list iterator will avoid this by remembering where it is.

The iterator will be a separate object from the list, with a pointer into the
list. Single links are shown for simplicity.

To do this, the `LinkedListIterator` class will need one instance variable, `next`,
that is a reference to where the next item in the list is. Its methods will be
(see the interface in Table 6.2):

next() will get the element from the node pointed to by `next`, move `next` to
point to the next node, and then return its result.

hasNext() will test that `next` is not null.

The `next` field starts at the first regular (non-dummy) node in the list.

Listing 6.3 shows how to put these pieces together into code that goes inside
the `LinkedList` class. The `LinkedListIterator` class defines this new iterator
object, and the public `iterator()` method simply creates and returns an iterator. Adding the public `iterator()` method to the `LinkedList` class allows
the `LinkedList` class to implement the `Iterable` interface (Table 6.3).

Listing 6.3: Linked List Iterator

```
1   import java.util.Iterator;
2
3   public Iterator<E> iterator() {
4       return new LinkedListIterator();
5   }
6
7   private class LinkedListIterator implements Iterator<E> {
8       private Node<E> next = head.next;   // skip dummy
9
10      public boolean hasNext() {
11          return next != null;
12      }
13
14      public E next() {
15          if (!hasNext()) throw new NoSuchElementException();
16          E result = next.data;
17          next = next.next;
18          return result;
19      }
20
21      public void remove() {
22          throw new UnsupportedOperationException();
23      }
24  }
```

Privacy Issues

It might seem strange that the iterator can have a pointer into the linked list like this. Part of what makes it work is that the iterator class is nested inside the linked list class; in other words, an outside class could not do this. The second thing to notice is that the iterator class is also private, so the outside world can only access the iterator through the public `iterator()` method. That method and the public interface of the iterator itself tightly control how the iterator can be used.

There is also a new mechanism being used here that gives the iterator access to the nodes in the linked list: the iterator class is not static. It is our first example of an inner class.

Inner Classes

Recall from Section 3.3 that our nested `Node` classes have been declared static because they do not need access to any of the instance variables of their containing class. You may have noticed that the nested class in Listing 6.3 is not static. An **inner class** is a non-static nested class, and it *does* have access to the instance variables of its containing object. In Listing 6.3, the iterator needs access to the head of the linked list, and so it must be an inner class rather than a static nested class.

Extending Interfaces

One other point will be useful for completing this implementation. Recall (see page 50) that a class may implement more than one interface, so we might indicate that `LinkedList` implements `Iterable` like this:

```
public class LinkedList<E> implements List<E>, Iterable<E> ...
```

However, it is reasonable to expect all lists to be iterable, and this can be done with **interface inheritance**. The **extends** keyword allows interfaces to inherit from each other in the same way as classes:

```
public interface SubInterface extends SuperInterface { ... }
```

The subinterface inherits all methods from the superinterface and then adds new methods of its own. Therefore, to require all lists to be iterable, we extend the interface:

```
public interface List<E> extends Iterable<E> { ... }
```

Using Iterators

The motivation for creating an iterator was the inefficiency of this loop for a linked list:

```
for (int i = 0; i < items.size(); i++) {
    System.out.println(items.get(i));
}
```

Rewriting the loop to use an iterator looks like this, assuming `items` is an Integer list:

```
Iterator<Integer> it = items.iterator();
while (it.hasNext()) {
    System.out.println(it.next());
}
```

This provides the $O(n)$ traversal we were seeking. It is a common enough task that Java provides the enhanced for-loop (see page 18) for even more convenience. The same loop can be written as:

```
for (Integer item : items) {
    System.out.println(item);
}
```

The enhanced **for** uses the iterator to get the same $O(n)$ performance while hiding all of the iterator details. All that is required to use it is that the object have an `iterator()` method, which is the same as implementing the `Iterable` interface (Table 6.3).

Exercises

1. Explain why a linked list iterator allows traversing the list in $O(n)$ time.

2. Modify the `List<E>` interface to extend `Iterable<E>`.

3. Incorporate the modifications in this section to the `LinkedList` class so that it implements the `Iterable` interface. Test the implementation by explicitly using the iterator and with an enhanced for-loop.

4. Without an iterator, an `ArrayList` already provides efficient traversals of its elements. Even so, describe an advantage of modifying `ArrayList` to support the `Iterable` interface.

5. Modify the `ArrayList` class from Section 6.2 to support the `Iterable` interface. Test the implementation with an enhanced for-loop. Hint: draw a diagram like the one on page 119 and follow the steps used there for linked lists.

Chapter 7

Recursion

7.1 Mathematical Functions

Consider the factorial function

$$n! = n \cdot (n-1) \cdot (n-2) \cdots 2 \cdot 1$$

The "$\cdots$" in this expression is a bit informal, even though we intuitively know what it means. One way to precisely define factorial looks like this:

$$n! = \begin{cases} 1 & \text{if } n = 0 \\ n \cdot (n-1)! & \text{if } n > 0 \end{cases}$$

If we use this definition to evaluate 3!, it unravels like this:

$$
\begin{aligned}
3! &= 3 \cdot 2! \\
&= 3 \cdot (2 \cdot 1!) \\
&= 3 \cdot (2 \cdot (1 \cdot 0!)) \\
&= 3 \cdot (2 \cdot (1 \cdot 1)) \qquad = 6
\end{aligned}
$$

Notice that the first part of the definition causes the unraveling to stop.

This definition is **recursive** because it defines factorial in terms of another factorial; in this case, $n!$ is defined in terms of $(n-1)!$. Recursive definitions have two key features:

Recursive steps use smaller arguments. In this example, the factorial of n is computed using the smaller argument $n - 1$.

Base cases cause recursion to stop. Here, the base case is that 0! is defined to be 1. There is never any recursion in a base case—that is what causes the recursion to stop. More than one base case may be necessary if the recursive step uses more than one smaller argument.

Recursive Functions in Java

Like most modern programming languages, Java supports recursion by allowing functions to call themselves. In fact, the mathematical definition of $n!$ above translates almost directly into the Java of Listing 7.1.

Listing 7.1: Recursive Factorial

```java
public class Recursion {
    public static int factorial(int n) {
        if (n <= 1) return 1;
        return n * factorial(n - 1);
    }
}
```

In code, base cases are normally tested first. Code may also take advantage of small efficiencies, such as returning 1 immediately for both 0! and 1!. Throughout this chapter we assume integer parameters n are nonnegative.

Tracing Recursive Calls

Tracing a recursive function call is like unraveling the mathematical definition, except that in this case we can stop one step earlier:

```
factorial(3)
       ↳ 3 * factorial(2)
               ↳ 2 * factorial(1)
                       ↳ ①
```

When `factorial(1)` returns 1, that allows the call to `factorial(2)` to return $2 * 1 = 2$, which then allows the call to `factorial(3)` to return $3 * 2 = 6$.

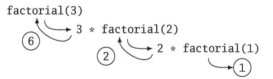

Natural Recursions

The following structures are natural candidates for recursive function definitions because they themselves can be defined recursively:

Natural numbers Recursive functions on natural numbers (the nonnegative integers) usually call themselves on $n - 1$ or smaller values, and stop at 0 or 1.

Strings Recursive functions on strings usually call themselves on substrings and stop on the empty string or null.

Lists Recursive functions on lists usually call themselves on sublists and stop on the empty list or null. This works naturally with linked lists because any given node has a smaller list that follows it via the **next** pointer.

Any object types in Java being used for recursion usually include a test for null as one of their base cases. Trees are another naturally recursive structure we will study beginning in Chapter 8.

Exercises

1. Use this recursive definition of add(m, n) to:

$$\text{add}(m, n) = \begin{cases} m & \text{if } n = 0 \\ 1 + \text{add}(m, n - 1) & \text{if } n > 0 \end{cases}$$

   (a) Trace the call **add(11, 3)**
   (b) Implement **add(m, n)** as a recursive Java function for integers **m** and **n**.

2. Use this recursive definition of mul(m, n) to:

$$\text{mul}(m, n) = \begin{cases} 0 & \text{if } n = 0 \\ m + \text{mul}(m, n - 1) & \text{if } n > 0 \end{cases}$$

   (a) Trace the call **mul(2, 4)**
   (b) Implement **mul(m, n)** as a recursive Java function for integers **m** and **n**.

3. Use this recursive definition of pow(m, n) to:

$$\text{pow}(m, n) = \begin{cases} 1 & \text{if } n = 0 \\ m \cdot \text{pow}(m, n - 1) & \text{if } n > 0 \end{cases}$$

   (a) Trace the call **pow(3, 3)**
   (b) Implement **pow(m, n)** as a recursive Java function for integers **m** and **n**.

4. Write a recursive Java function **length(String s)** that computes the length of the string **s**.

5. Write a recursive Java function **reverse(String s)** that returns the string **s** in reverse.

6. Write a recursive Java function **fib(n)** that computes the nth term in the **Fibonacci sequence**

$$1, 1, 2, 3, 5, 8, 13, 21, 34, \ldots$$

   where each term is the sum of the two before it.

7. Write a recursive Java function to implement Euclid's algorithm for computing the greatest common divisor `gcd(m, n)`, which is the largest integer `k` dividing both `m` and `n`. (See also Exercise 8 in Section 1.1.) The recursive version of Euclid's algorithm is based on these two facts:

$$\gcd(m, n) = \gcd(n, m \bmod n)$$
$$\gcd(m, 0) = m$$

Notice that $m \bmod n$ is always less than n.

8. Write a recursive Java function `bits(n)` that returns a string representing the nonnegative integer `n` in binary. Hint: the last bit is `n % 2`, and you can get the other bits with recursion on `n / 2`. You may need more than one base case.

7.2   Visualizing Recursion

In this section, we look at two techniques for visualizing recursion that can help you follow how recursive functions work: call trees and the call stack. The call stack will also begin to explain how function calls are implemented in general.

Recursive Call Trees

A **call tree** diagrams the set of calls that are generated by a recursive function call. It generalizes the idea of tracing a recursive function described in the previous section by not following return values or calculations but only tracking the different function calls. Call trees are also often drawn for n rather than tracing a specific value, in order to understand the overall behavior of the function. The point of a call tree is usually to count the total number of function calls.

For simple recursive functions, the call tree is a single chain of calls. For example, the tree for `factorial(n)` is:

```
factorial(n)
        ↳ factorial(n - 1)
                ↳ factorial(n - 2)
                        ↳ ...
                            ↳ factorial(1)
```

This indicates that a recursive call to `factorial(n)` generates approximately n function calls.

If a recursive function calls itself more than once, a **tree** of calls results. For example, calling the recursive Fibonacci function `fib(5)` from Exercise 6 in the previous section generates this tree:

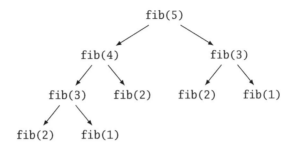

Analyzing Recursive Functions

The call tree gives some intuition for the time complexity of a recursive function: for example, in the case of the Fibonacci function, a lot of calls means a lot of work!

In simple cases, if each function call has $O(1)$ steps, as in the case of the factorial function, then counting the function calls gives us the total amount of work. For the factorial function, since each call is $O(1)$ and the tree generates n calls, the total work is $O(n)$.

More advanced techniques for analyzing recursive functions using recurrence relations are studied in later courses.

Implementing Recursion

Imagine the call tree for `fib(100)`, or even `fib(10)`. How does a recursive function keep track of where it is as it works through all of those function calls? As functions return, where do their return values go? The answers to these questions lie in how functions calls are implemented in most languages.

Function Call Stack

Every time a function call is made, a section of memory called a **stack frame** is pushed on to the **call stack**. The call stack is just a regular stack that happens to hold stack frames; it is also known as the **run-time stack**, and the frames are sometimes called **activation records**.

A function is able to use its stack frame memory while it is active, and then when the function call finishes, its stack frame is popped. Usually, the return value of the function (if any) will be put in a special register or left at the top of the stack.

In general, stack frames contain the following basic information:

Local Variables
Return Address
Argument Values

The caller pushes the argument values and return address, and then the function being called pushes space for its local variables. When the function that was called finishes, the frame is popped and the return address is used to jump back to where it had been called from. This is how a function knows "where it is," in the sense asked above.

This general method for implementing function calls makes it so that nothing special is required to support recursion. When a function calls itself, it simply pushes a new frame on the stack with the appropriate values. There is generally a limited amount of memory available for the stack, however. Any recursion without a base case (or simply generating too many calls) will result in a **stack overflow error**.

Visualizing the Call Stack

The factorial method has no local variables, and so we can trace a call to `factorial(3)` with this sequence of views of the call stack, showing only the argument value of n inside each frame and return values left on the top of the stack:

The final value 6 is left on the top of the stack for whatever method called `factorial(3)`.

Stack Traces

The Java run-time environment will print a **stack trace** after an exception, to allow you to see how the problem occurred. A stack trace simply shows the functions currently on the call stack, beginning at the top. Many debuggers also allow you to view the call stack during program execution.

Exercises

1. Draw a call tree for the add(m, n) function in Exercise 1 of the previous section. Use the tree to determine the $O()$ performance of the function.

2. Draw a call tree for the mul(m, n) function in Exercise 2 of the previous section. Use the tree to determine the $O()$ performance of the function.

3. Draw a call tree for the pow(m, n) function in Exercise 3 of the previous section. Use the tree to determine the $O()$ performance of the function.

4. Write a recursive version of pow(m, n) to compute m^n for n $\geq$ 0 that is $O(\log n)$.

5. Count the number of times fib(1) and fib(2) appear in the call tree of fib(100). Hint: draw the top of the tree, and count the number of times each of fib(100), fib(99), fib(98), ... appear. Look for a pattern; the pattern changes slightly at n = 1. To see that change, look again at the tree for fib(5).

6. Show how the call stack changes during the evaluation of factorial(4).

7. Show how the call stack changes as a result of calling these recursive functions from the exercises in the previous section:

   (a) add(7, 3)

   (b) mul(5, 4)

   (c) pow(2, 3)

   (d) length("abc")

   (e) reverse("Java")

   (f) gcd(10, 6)

7.3   Recursive and Generalized Searches

Both linear and binary searches are good candidates for recursion. They each check one element, and if it is not the target, continue searching on a smaller list. The same process is used on the smaller list, and so it may be accomplished by calling the same recursive method.

Linear search from Section 1.3 and binary search in Section 2.3 were also both only designed for **int** arrays. As we revisit them, we generalize these searches to work with other types.

Public Wrapper Methods

There is really only one new concept needed to write recursive versions of linear and binary search. Recursive functions work by calling themselves on smaller parameters, but the parameters to linear and binary search leave no room for that:

```
linearSearch(data, target)
binarySearch(data, target)
```

Furthermore, the users of these search methods should not know or care (unless it affects performance) whether these methods are implemented by recursion or not, and so we do not want to change the parameter lists of the public methods.

The solution is to use the public methods as **wrappers** that call private recursive methods with additional parameters to drive their recursion:

Public wrapper `linearSearch(data, target)`
 ↳ Private recursive `linearSearch(data, target, left)`

Public wrapper `binarySearch(data, target)`
 ↳ Private recursive `binarySearch(data, target, left, right)`

The new parameters in the private methods control the section of the array currently being searched, which is what needs to change during the recursion. Binary search needs two parameters because it can work in from both ends, whereas linear search only searches from one end.

All the wrapper methods do is get the recursion started with initial values for the new parameters.

Generalized Linear Search: Objects

The linear search in Listing 1.3 works only with **int** arrays. To generalize it, we only need to be able to tell if the target equals an array item. Because the `equals()` method is defined in the `Object` class (see page 23), we can define linear search on the `Object` type and then it will work for arrays using any object type.

Generalized Binary Search: Comparables

Binary search requires more than testing for equality: we need to be able to compare elements with respect to an ordering that works like less than or greater than. Recall that Java strings have a `compareTo()` method (see Table 1.5) that makes exactly this type of comparison by returning a positive, negative, or zero integer value. Any type with such a `compareTo()` method implements the generic Java `Comparable<T>` interface in Table 7.1.

TABLE 7.1: Comparable Interface

`int compareTo(T other)`

Returns negative if this object is considered less than `other`, zero if equal, and positive if greater.

For example, the `String` class implements `Comparable<String>`. In general, classes that implement `Comparable` should also override `equals()` in a way that is consistent with `compareTo()`.

The generalized binary search is designed to work with any type implementing the `Comparable` interface.

Generic Methods

Because the `Comparable` interface is generic, to write a generalized binary search using that interface, we need to know how to declare a method with a generic type in it. All of the other generic methods we have written have been inside generic classes.

Java allows methods to introduce a generic type parameter; this is called declaring a **generic method**. The type variables go in between the method modifiers and the return type, like this:

```
modifiers <T> returnType name(parameters) { ... }
```

This allows the type parameter to be used anywhere in the rest of the method definition. For example,

public static `<T>` **int** `binarySearch(T[] data, T target) { ... }`

defines `binarySearch()` so that the array type has to match the target type.

We still need to specify that elements of type `T` are comparable. The rough idea is expressed like this:

`<T` **extends** `Comparable<T>>`

This says that objects of type `T` implement the `Comparable` interface. However, because of inheritance, that requirement is too restrictive. The right declaration is:

```
<T extends Comparable<? super T>>
```

See Chapter 11 of *The Java Programming Language* [1] if you are interested in the details.

Listing 7.2 shows how all of these ideas work together, providing a complete implementation of recursive binary search as a generic method with public wrapper.

Listing 7.2: Generalized Recursive Binary Search

```java
public static <T extends Comparable<? super T>> int
        binarySearch(T[] data, T target) {
  return binarySearch(data, target, 0, data.length - 1);
}

private static <T extends Comparable<? super T>> int
        binarySearch(T[] data, T target, int low, int high) {
  if (low > high) return -1;
  int mid = (low + high) / 2;
  int result = target.compareTo(data[mid]);
  if (result == 0) {
    return mid;
  } else if (result < 0) {
    return binarySearch(data, target, low, mid - 1);
  } else {
    return binarySearch(data, target, mid + 1, high);
  }
}
```

Integer[] and int[]

Finally, because these generalized methods are written for object types, they require arrays of objects rather than arrays of primitive types. In other words, the new versions of linear and binary search will not work on **int** arrays. Java automatically boxes and unboxes individual elements, but it will not automatically convert between **int**[] arrays and Integer[] arrays.

Exercises

1. Suppose the Integer array data contains

   8, 28, 31, 35, 39, 40, 44, 51

   Draw the call tree for each of these calls to the private recursive binary search function in Listing 7.2:

   (a) binarySearch(data, 32, 0, 7)

   (b) binarySearch(data, 51, 0, 7)

   (c) binarySearch(data, 5, 0, 7)

2. Implement recursive linear search on Object types.

3. Write a `main()` method to test recursive linear search from Exercise 2 on an array of:

   (a) Integers
   (b) Doubles
   (c) Strings

4. Write a `main()` method to test `binarySearch()` from Listing 7.2 on an array of:

   (a) Integers
   (b) Doubles
   (c) Strings

5. Determine the $O()$ performance of recursive linear search.

6. Determine the $O()$ performance of recursive binary search.

7. Generalize insertion sort (not recursively) from Listing 2.1 to work on any comparable type.

8. Generalize selection sort (not recursively) from Exercise 13 in Section 2.2 to work on any comparable type.

9. Modify the `Fraction` class from Listing 1.5 to implement the `Comparable` interface. Assume all denominators are positive for simplicity.

7.4 Applications

The problems we solved with recursion to this point all have simple non-recursive solutions that may even be preferable to their recursive counterparts. The point of those examples was to get comfortable with recursion without too much complication. The three problems in this section use recursion for their solution in more substantial ways. In fact, you might think about how to solve them without recursion.

Longest Common Subsequence

Given a string `s`, a **subsequence** of `s` is any string obtained by deleting any number of characters from `s` (including none). For example, `"acf"` is a subsequence of `"abcdef"` because the characters `b`, `d`, and `e` were deleted. Any substring is also a subsequence, but as this example shows, subsequences do not have to consist of consecutive characters.

The **longest common subsequence** problem asks to find the longest string lcs(s1, s2) that is a subsequence of both s1 and s2. For example,

   lcs("xayzzbxyczxxdyxx", "ttsatbstcsstdt")

is the string "abcd". This problem arises naturally in genetics when determining how similar two strands of DNA are.

A recursive lcs(s1, s2) is described in Listing 7.3. This recursive solution is very inefficient. However, it is the basis of a much more efficient solution using **dynamic programming** that you may see in a later algorithms course.

Listing 7.3: Longest Common Subsequence (Pseudocode)

```
1  If s1 or s2 is empty, return the empty string
2  If first characters match
3      Return matching character + lcs(rest of s1, rest of s2)
4  Else
5      Return longer of lcs(s1, rest of s2) and lcs(rest of s1, s2)
```

Towers of Hanoi

The **Towers of Hanoi** is a puzzle with three posts and a stack of disks on one of the posts that decrease in radius as they go up.

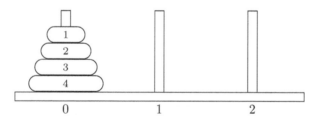

The object is to move all of the disks from post 0 to post 2 following these rules: (1) only one disk may be moved from one post to another per turn, and (2) a disk may never be moved onto a smaller disk.

The key insight for solving this puzzle is that if we could recursively move disks 1–3 to post 1, then disk 4 could move to post 2. A second recursive call could then finish by moving disks 1–3 from post 1 to post 2. To make this work, both the number of disks and the roles of each post need to be able to change for each recursive call. That means the recursive function needs parameters for the number of disks, as well as which post is used for each role: source, intermediate, and destination.

Listing 7.4 outlines a recursive hanoi(n, src, intermed, dest) based on these ideas.

Listing 7.4: Towers of Hanoi (Pseudocode)

```
1  Move n-1 disks from src to intermed using dest
2  Move disk n from src to dest
3  Move n-1 disks from intermed to dest using src
```

Backtracking

Problems that involve searching a large number of possibilities can make it difficult to keep track of where an algorithm "is" currently in the search space. Example problems of this type include mazes, puzzles such as crosswords or Sudoku, and discrete optimization problems that attempt to maximize or minimize some quantity over a set of discrete objects. Recursive **backtracking** solves problems like these by using the call stack to manage where it is and to allow backing up to try something else if the current search path has failed.

The *n*-**queens problem** is a puzzle that requires placing n chess queens on an $n \times n$ board so that none of the queens attack each other. Queens are allowed to move horizontally, vertically, or diagonally any number of squares. For example, with $n = 4$, the following is not a valid solution because two queens attack each other along a diagonal:

Q			
		Q	
			Q
	Q		

A recursive backtracking solution to the *n*-queens problem reduces the problem to placing one queen in each column. Listing 7.5 describes a recursive `queens(board, column)` function that moves across the board from left to right, placing one queen per column.

Listing 7.5: Backtracking Solution To *n*-Queens (Pseudocode)

```
1  If finished, display board
2  Else
3      For each row
4          board[column] = row
5          If this is a legal position
6              queens(board, column + 1)
```

Where is the backtracking in this code? It is managed completely by the recursion: if `queens(board, column + 1)` fails to find a solution, the for-loop advances to check the next possible row.

Exercises

1. Implement Listing 7.3 as a recursive function `lcs(s1, s2)` to return the longest common subsequence of strings `s1` and `s2`.

2. Implement Listing 7.4 as a recursive `hanoi(n, src, intermed, dest)` to solve the Towers of Hanoi puzzle. Print a message like this for the middle step of the method:

   ```
   Move disk 3 from 0 to 2.
   ```

   Do not forget a base case.

3. Determine the $O()$ performance of the `hanoi()` function by counting the number of function calls made with n disks.

4. Implement Listing 7.5 as a recursive function `queens(board, column)` to solve the n-queens problem using backtracking.

5. Explain why the recursive backtracking solution to the n-queens problem finds all solutions to the problem, rather than just one.

Chapter 8

Trees

Trees are non-linear, hierarchical, and recursive data structures with a wide range of applications. In fact, we have already seen call trees in Section 7.2 as a useful model for tracking recursive computations.

8.1 Definitions and Examples

A **tree** is a (possibly empty) set of elements called **nodes** with one node designated as the **root**. The root is connected via **edges** to some number of **child** nodes, each of which is itself the root of a **subtree**. Thus, trees are naturally recursive. A **leaf** is a node with no children; an **internal node** is any non-leaf. A **branch** in a tree is a path of connected edges starting at the root and ending in a leaf. Trees are usually drawn with the root at the top.

For example, in the tree below, 31 is the root, 7 is a child of 5, 24 is a leaf, and 14 is an internal node.

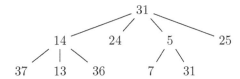

The child relationship between nodes in a tree naturally leads to many other familial terms for nodes, such as **parent, grandparent, sibling, ancestor,** and **descendant**. Every node has exactly one parent, except for the root which has no parent.

The **depth** or **level** of a node n in a tree is defined recursively:

$$\text{depth(n)} = \begin{cases} 1 & \text{if n is the root} \\ 1 + \text{depth(parent(n))} & \text{otherwise} \end{cases}$$

The **height** of a tree is its maximum depth. The tree above has height 3, and node 25 is on level 2.

Among the most common trees are **binary trees** in which every node has at most two children, usually named **left** and **right**. The following examples illustrate just a few of the many uses of binary trees.

Expression Trees

Arithmetic expressions can be represented as binary trees by translating each binary operation to a tree with its operator as the root and its two operands as the operator's children. For example, $(a + b) * c$ can be represented as this binary tree:

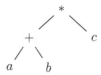

Such a tree is called an **expression tree** or **parse tree**, because it represents the syntactic structure of the expression.

There is a natural correspondence between subexpressions and subtrees in expression trees. For example, $(a + b)$ is a subexpression of $(a + b) * c$, and it occurs as the left subtree of the expression tree, since it is the left operand of the multiplication:

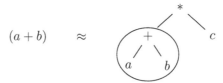

Thus, when drawing expression trees, it may be easier to start with subexpressions and build the tree from the bottom up rather than trying to work from the top down.

Example

The expression tree corresponding to $(a - c) * d + \dfrac{f}{b + e}$ is:

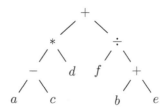

Notice the subtrees corresponding to subexpressions such as $(a - c)$, $(a - c) * d$, and $b + e$.

Binary Search Trees

A **binary search tree** is a binary tree containing comparable values in which the following property is true at every node **p**:

$$\begin{matrix} \text{Every value in} \\ \text{left subtree of } \mathbf{p} \end{matrix} \; < \; \text{Value at } \mathbf{p} \; < \; \begin{matrix} \text{Every value in} \\ \text{right subtree of } \mathbf{p} \end{matrix}$$

For example, here is a small binary search tree:

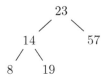

The inequalities above are strict because we assume binary search trees do not contain duplicate elements; see Exercise 6 in Section 9.4 for the reason. We will study binary search trees in depth in Chapter 9.

Representing General Trees as Binary Trees

Requiring every node in a tree to have at most two children might sound like a serious limitation compared with allowing any number of children at any node. At the same time, writing code that will store any number of children will be far more complicated than nodes with two children, left and right.

Are general trees "more powerful" than binary trees in the sense that they can store relationships that binary trees cannot? The answer is no: any tree can be stored as a binary tree and converted back.

The **left child/right sibling** convention takes an arbitrary tree and has each node point to its first child (if any) on the left, and its next sibling (if any) on the right. For example, given this general tree:

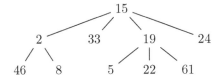

the associated binary tree begins with 15 at the root and 2 on its left as its first child. As the root, 15 has no siblings so has no right child:[1]

---

[1]Siblings of the root could be considered other roots in the same forest. This allows an entire forest of general trees to be represented by one binary tree.

Then 2 has 46 on the left as its first child and 33 on the right as its next sibling:

The entire associated binary tree is:

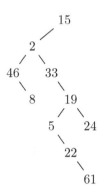

Exercises

1. In a binary expression tree, what type of node are operators always in?

2. In a binary expression tree, what type of node are operands always in?

3. Draw a binary expression tree for each of these arithmetic expressions:

   (a) $x + y + z * w - \dfrac{v}{u}$

   (b) $\dfrac{a * b}{c} - d + f * g + h$

   (c) $(x + y + z) * \dfrac{w - v}{u}$

   (d) $\dfrac{x * y}{z - w + v * u}$

4. For each of these lists of elements, draw three binary search trees of different heights containing exactly these items:

   (a) 19, 27, 38, 42, 48, 49, 62

   (b) 6, 10, 61, 77, 81, 84, 94

   (c) 27, 48, 56, 61, 63, 76, 89

5. Determine the maximum height of a binary search tree with n nodes.

6. Determine the minimum height of a binary search tree with n nodes.

7. Draw the associated binary tree for each of these general trees using the left child/right sibling convention:

(a)

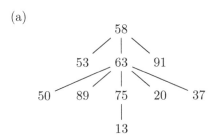

(b)

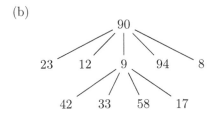

8. Draw the associated general tree for each of these binary trees using the left child/right sibling convention:

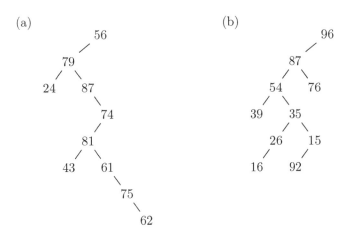

9. Draw a binary tree with five nodes that is its own associated general tree using the left child/right sibling convention.

8.2   Traversals

Recall that a tree has a root and then some number of children, each of which is itself the root of a subtree. This recursive structure—that trees are built from subcomponents that are also trees—leads to recursive traversal methods for binary trees. As with linked lists (see Section 6.3), a **traversal** of a tree is a method that visits every node in the tree.

Depth-First

If the tree is binary, then there are three ways to naturally order the steps of a recursive traversal, depending on when the node itself is visited:

Preorder (node, left, right) visits the node, traverses the left subtree, and traverses the right subtree.

Inorder (left, node, right) traverses the left subtree, visits the node, and traverses the right subtree.

Postorder (left, right, node) traverses the left subtree, traverses the right subtree, and visits the node.

Visiting a node just refers to whatever a method needs to do with each node, such as printing, counting, or modifying the node.

Each of these traversals is **depth-first** because it traverses all the way down some branch before backtracking to reach other nodes. Depth-first traversals may be defined for other trees, as well, but we limit our use to binary trees.

Example

Consider this binary tree:

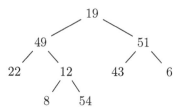

By hand, the preorder traversal is probably easiest because nodes are visited first as you reach each subtree:

19		—————19L———			—19R—			
19	49	49L	—49R—		—19R—			
19	49	22	—49R—		—19R—			
19	49	22	12	8	54	—19R—		
19	49	22	12	8	54	51	43	6

Inorder traversal, on the other hand, benefits from a top-down view:

	———19L———				19	——19R——		
49L	49	——49R——			19	51L	51	51R
22	49	12L	12	12R	19	43	51	6
22	49	8	12	54	19	43	51	6

If you write inorder traversals from beginning to end rather than top-down, then you need to keep track of when subtrees are finished. For example, after node 54, when the left subtree of 19 is finished, then you need to remember to visit the 19 node. Postorder is similar to inorder except that the node itself comes last:

	————19L————				——19R——			19
49L	——49R——			49	51L	51R	51	19
22	12L	12R	12	49	43	6	51	19
22	8	54	12	49	43	6	51	19

Breadth-First

A **breadth-first** traversal visits nodes in **level order**: level 1 first, then level 2, and so on. Like preorder traversals, the root will always be visited first in a breadth-first traversal. Breadth-first traversals are not naturally recursive. In fact, relative to the connections that exist between parents and children, a breadth-first traversal appears to jump around the tree.

Using a queue to keep track of the next nodes to visit leads to the nice algorithm in Listing 8.1. Because any number of children may be enqueued in step 4, the breadth-first traversal algorithm may be used with any tree.

Listing 8.1: Breadth-First Traversal (Pseudocode)

```
1  Enqueue root
2  While queue is not empty
3      Dequeue and visit node
4      Enqueue children of node
```

Example

Breadth-first traversals are easy by hand because you just trace across the levels. To see how the queue works in Listing 8.1, consider this small binary tree:

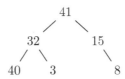

Then the sequence of operations looks like this:

Dequeue and Visit	Enqueue	Queue Contents
	41	41
41	32, 15	32, 15
32	40, 3	15, 40, 3
15	8	40, 3, 8
40		3, 8
3		8
8		

Exercises

1. Describe the base case for the three recursive depth-first traversals.

2. List the order in which nodes are visited using each of these traversals on the tree below:

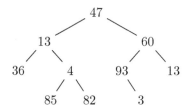

  (a) Preorder

  (b) Inorder

  (c) Postorder

  (d) Breadth-first   Show how the queue contents change using List-ing 8.1, as in the example above.

3. List the order in which nodes are visited using each of these traversals on the tree below:

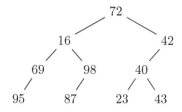

  (a) Preorder

  (b) Inorder

  (c) Postorder

  (d) Breadth-first   Show how the queue contents change using List-ing 8.1, as in the example on page 143.

4. Give preorder, postorder, and inorder traversals of your answers to Exercise 3 from Section 8.1.

5. Describe the effect of each of the depth-first traversals on a binary expression tree.

6. Draw a binary expression tree corresponding to each of these:

    (a) Postorder traversal is: a b c d e * + / –

    (b) Postorder traversal is: a b * c + d e / –

    (c) Preorder traversal is: / – a b * + c d e

    (d) Preorder traversal is: + – a * b c / d e

7. Give preorder, postorder, and inorder traversals of your answers to Exercise 4 from Section 8.1.

8. Describe the effect of an inorder traversal on a binary search tree.

9. Draw a binary search tree corresponding to each of these:

    (a) Breadth-first traversal is: 142 41 191 752 532 787 408 743 962 373

    (b) Breadth-first traversal is: 81 26 561 509 921 265 586 899 693 779

    (c) Breadth-first traversal is: 654 209 833 208 318 679 99 521 422 461

10. Listing 8.1 shows how a queue is used for a breadth-first traversal. Is there a natural role for a stack in the depth-first traversals? If so, describe it; if not, explain why not.

8.3   Binary Tree Abstract Class

As a prelude to implementing binary search trees in Chapter 9, this section develops a general `BinaryTree` class to represent any binary tree. It defines the basic structure common to any binary tree: a root, a node structure, and traversal methods. However, it will not be used for creating objects. It is designed only to be extended by more specific classes like `BinarySearchTree`.

Table 8.1 gives the public API of the `BinaryTree` class. (Since `BinaryTree` is a class and not an interface, we are not considering it an ADT.) The `toString()` method will also be overridden to return one of these traversals.

TABLE 8.1:   Binary Tree API

`String breadthFirst()`
Returns breadth-first traversal of this tree.
`String inorder()`
Returns preorder traversal of this tree.
`String preorder()`
Returns preorder traversal of this tree.
`String postorder()`
Returns preorder traversal of this tree.

Abstract Classes

Java **abstract classes** are intended for exactly this purpose: providing part of a class implementation but not a complete one. An abstract class may not be used to create objects; it can only be extended via inheritance. Objects are created by **concrete classes** that extend the abstract class and are not abstract themselves.

The syntax for declaring an abstract class is to include the **abstract** keyword as a modifier in the class definition:

```
public abstract class AbstractClass { ... }
```

Abstract classes may be generic and include type parameters. They may define fields, methods, and constructors, although constructors are usually declared protected because they are not meant to be called directly. In fact, protected visibility is common in abstract classes because they are designed to be extended.

Binary Tree Class

Listing 8.2 begins the `BinaryTree` implementation as an abstract class. Notice its use of all three levels of visibility. The root and `Node` class (along with the components of the `Node` class) are protected so they can be used by subclasses. The `preorder()` method is part of the public interface and acts as a wrapper to call a private recursive function. Not even subclasses have a need to know about the private recursive function.

Also notice how `toString()` is used in Listing 8.2. The `Node` class overrides `toString()`, returning `data.toString()` on line 24. Once this method has been written, nodes can be inserted into string expressions with ease, as on line 10 of the recursive `preorder()` method. This allows other methods in the `BinaryTree` class and its subclasses to think in terms of nodes rather than their components when building strings.

The exercises ask you to finish the implementation.

Listing 8.2: Binary Tree

```
1   public abstract class BinaryTree<E> {
2       protected Node<E> root;
3
4       public String preorder() {
5           return preorder(root);
6       }
7
8       private String preorder(Node<E> n) {
9           if (n == null) return "";
10          return (n + " " + preorder(n.left) + " " +
11                  preorder(n.right)).trim();
12      }
13
14      protected static class Node<T> {
15          protected T data;
16          protected Node<T> left, right, parent;
17
18          protected Node(T data, Node<T> parent) {
19              this.data = data;
20              this.parent = parent;
21          }
22
23          public String toString() {
24              return data.toString();
25          }
26      }
27  }
```

Exercises

1. Add these public methods to the `BinaryTree` class in Listing 8.2:

   (a) `inorder()`   Use a private recursive traversal.

   (b) `postorder()`   Use a private recursive traversal.

   (c) `toString()`   Call one of the traversals.

2. Write a public `breadthFirst()` traversal method for the `BinaryTree` class in Listing 8.2. Implement the algorithm from Listing 8.1, using your own queue class and a `StringBuilder` to accumulate the results. Set up the queue to store `Node<E>` objects, and be careful not to put null entries in the queue.

3. The call to `trim()` in line 11 of Listing 8.2 cleans up some extra spaces, but it can still leave different numbers of internal spaces between items in preorder and postorder traversals. Fix this problem so that there is always one space between elements in the traversals.

Project: A Collection Hierarchy

You may have noticed some code duplication with the stack, queue, and list data structures. Abstract classes can help reduce that duplication by defining common fields and methods. There is no one best way to decide what goes into an abstract superclass, but the idea is to aim for a design that minimizes duplicate code.

If we map out what was common among stacks, queues, and lists, it looks something like this:

All of the structures had `size()` and `isEmpty()` methods. Some also included a `size` instance variable.

Array structures all had a `data` array instance variable, two constructors, and a `resize()` method. The queue needed its own specialized version of `resize()` and had a separate `capacity` instance variable.

Linked structures defined a private node class and had at least a `head` instance variable.

These common features suggest the following inheritance hierarchy:

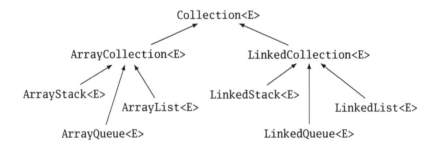

Remember that in addition to inheriting fields and methods from their superclass, subclasses may define new fields and methods, as well as override existing methods.

Exercises

1. Develop a Collection hierarchy as described above. Write the necessary abstract classes and modify the concrete classes to participate in the hierarchy.

   Although a general approach is suggested, you will still need to make specific design decisions along the way. Write a short report explaining the choices you make, including how you approached those decisions and the alternatives you considered.

Chapter 9

Binary Search Trees

The next several chapters focus on data structures that provide fast search, insertion, and deletion. Binary search trees generally give very good performance on all three of these operations. Recall from Section 8.1 that we assume binary search trees do not contain duplicate elements.

Table 9.1 lists the public API of the the `BinarySearchTree` class.

TABLE 9.1: Binary Search Tree API

void add(E item) Adds item to tree if not already present.
boolean contains(E item) True if item is in tree.
E min() Returns smallest item in tree.
E max() Returns largest item in tree.
E pred(E item) Returns inorder predecessor of item in tree.
boolean remove(E item) True if item is found and removed from tree.
E succ(E item) Returns inorder successor of item in tree.

We begin with methods that query the tree.

9.1 Queries

Binary search trees are built for fast searching using an algorithm just like binary search from Section 2.3. That algorithm required a sorted list, but a binary search tree can be considered sorted if we use an inorder traversal. Unlike a sorted list, however, many different binary search trees can be built from the same set of elements.

Search

The definition of binary search tree on page 139 is, as its name implies, built for binary search:

$$\text{Every value in left subtree of } \mathbf{p} \; < \; \text{Value at } \mathbf{p} \; < \; \text{Every value in right subtree of } \mathbf{p}$$

The recursive algorithm for searching a binary search tree is based on comparing the search key to a node's data value, beginning at the root. If the key matches the node's data value, it is found; otherwise, search in the left subtree if the key is smaller than the node's value or search in the right subtree if it is larger. If the current node is ever null, then the item is not in the tree.

Listing 9.1 begins the `BinarySearchTree` class with the `contains()` method and a private recursive `findNode()`. All binary search tree methods will be written in the same way, with a public wrapper calling a private node-based method, which may or may not be recursive. The `findNode()` method will also be useful for other methods. Notice its similarity to the original binary search in Listing 2.3 and recursive binary search in Listing 7.2.

Listing 9.1: Binary Search Tree Search

```java
public class BinarySearchTree<E extends Comparable<? super E>>
        extends BinaryTree<E> {
    public boolean contains(E item) {
        return findNode(item, root) != null;
    }

    private Node<E> findNode(E item, Node<E> n) {
        if (item == null || n == null) return null;
        int result = item.compareTo(n.data);
        if (result == 0) {
            return n;
        } else if (result < 0) {
            return findNode(item, n.left);
        } else {
            return findNode(item, n.right);
        }
    }
}
```

The `BinarySearchTree` class extends the `BinaryTree` class from Section 8.3. And as in Section 7.3, the type parameter E needs to be declared as

   `<E extends Comparable<? super E>>`

to guarantee that elements are comparable.

Minimum and Maximum

The structure of a binary search tree allows easy access to the largest and smallest elements in the tree. The smallest element is found by following left links as far as possible, and the largest is found by following right links as far as possible:

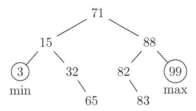

Thus, the minimum of this tree is 3 and the maximum is 99. These algorithms can also be applied to any subtree to find the largest and smallest elements of the subtree by simply beginning at the root of the subtree. For example, going right as far as possible from 15 leads to 65, which is the maximum of the subtree rooted at 15.

Predecessor and Successor

The **predecessor** of an element in a binary search tree is the element that precedes it during an inorder traversal. The **successor** is defined similarly as the element immediately following. The minimum in a tree has no predecessor, and the maximum has no successor.

The predecessor is easy to find if the node has a left subtree: it is the maximum of the left subtree. For example, the predecessor of 71 in the tree above is 65 because, as we found, 65 is the maximum of the left subtree rooted at 15.

However, if there is no left subtree, we need to move *up* the tree to find the predecessor. Consider the node 82 in the same tree. The predecessor of 82 is found by moving up toward the root using parent links until a right-child relationship is found. In this example, the link between 82 and 88 is a left-child relationship, but the link between 71 and 88 is a right-child. Thus, 71 is the predecessor of 82:

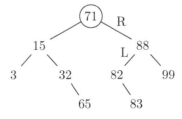

Listing 9.2 gives the code for both the public **pred()** method and its private

helper `predNode()`. Notice that `pred()` uses `findNode()` to get started, and then `predNode()` uses a private `maxNode()` helper (see Exercise 5b) for the case when `n.left` is not null.

Listing 9.2: Binary Search Tree Predecessor

```java
 1   public E pred(E item) {
 2       Node<E> n = findNode(item, root);
 3       if (n == null) return null;
 4       Node<E> pred = predNode(n);
 5       return (pred != null) ? pred.data : null;
 6   }
 7
 8   private Node<E> predNode(Node<E> n) {
 9       if (n.left != null) return maxNode(n.left);
10       Node<E> p = n.parent;
11       while (p != null && p.right != n) {
12           n = p;
13           p = n.parent;
14       }
15       return p;
16   }
```

Conditional Operator

Line 5 of Listing 9.2 uses the Java **conditional operator**. The conditional operator "?:" creates an expression via this syntax:

```
test ? valueIfTrue : valueIfFalse
```

This expression has the value `valueIfTrue` if the boolean `test` is true; otherwise, it has the value `valueIfFalse`. For example, the expression in line 5:

```java
    return (pred != null) ? pred.data : null;
```

returns `pred.data` if `pred` is not null, and returns null otherwise. It is exactly equivalent to the longer

```java
    if (pred != null) {
        return pred.data;
    } else {
        return null;
    }
```

Conditional expressions are useful for assignments or return statements like this that involve a quick test. Parentheses around the test are optional, but those in Listing 9.2 mimic the required parentheses in an if-statement.

Performance

All of the above queries in a binary search tree are $O(h)$, where h is the height of the tree. We postpone analyzing these operations in terms of n, the number of nodes in the tree, until Section 9.4.

Exercises

1. Use this binary search tree to answer the questions below.

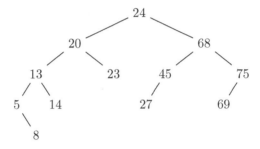

   (a) List nodes in the order they are visited in a search for the key 14.

   (b) List nodes in the order they are visited in a search for the key 51.

   (c) List nodes in the order they are visited in a search for the key 22.

   (d) List nodes in the order they are visited in a search for the key 69.

   (e) Determine the minimum and maximum elements in the tree.

   (f) Determine the predecessor and successor of 13 in the tree.

   (g) Determine the predecessor and successor of 24 in the tree.

   (h) Determine the predecessor and successor of 27 in the tree.

   (i) Determine the predecessor and successor of 68 in the tree.

2. Give one reason the binary tree Node class in Listing 8.2 includes a parent reference.

3. Describe the base case(s) for the recursion in the findNode() method of Listing 9.1.

4. Could the predNode() method of Listing 9.2 be written recursively? Explain why or why not.

5. Implement the following methods in the `BinarySearchTree` class:

   (a) `min()`   Write a private `minNode(n)` that returns the node with the smallest item in the subtree rooted at n.

   (b) `max()`   Write a private `maxNode(n)` that returns the node with the largest item in the subtree rooted at n.

   (c) `succ()`   Write a private `succNode(n)` that returns the node containing the successor of **n**.

   Note: `minNode()` is needed for `succNode()`.

   (d) `toString()`   Override the `BinaryTree` definition to force an inorder traversal.

6. If your solution to Exercise 5a used a recursive `minNode()`, write a non-recursive version, or vice versa.

7. If your solution to Exercise 5b used a recursive `maxNode()`, write a non-recursive version, or vice versa.

8. Explain why each of these binary search tree queries is $O(h)$, where h is the height of the tree:

   (a) Search
   (b) Minimum
   (c) Maximum
   (d) Predecessor
   (e) Successor

9.2   Insertion

Inserting into a binary search tree follows the same pattern as search. In fact, an unsuccessful search ends at a null pointer that is exactly where the item should be inserted. This idea is the basis for the `add()` method of the `BinarySearchTree` class.

For example, to insert 24 in this binary search tree, the search path ends left of 26:

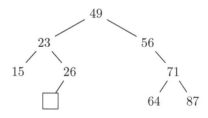

As with queries from the previous section, the public method `add()` will call a private `addNode()`.

The `add()` method handles a null item or changing the root. It should do the following:

```
If the item to insert is null, do nothing
If the root is null, insert item at root
Else start addNode at the root
```

The private `addNode()` method is outlined in Listing 9.3. It has the same structure as `findNode()`, except that it checks for a null child before making its recursive calls. If the child in the correct direction is null, that is the place to insert a new node.

Unlike `findNode()` and `predNode()`, `addNode()` is a **void** method because its purpose is to insert a new node rather than to return a particular node. You will be asked to implement this algorithm in the exercises.

Listing 9.3: Binary Search Tree Insertion (Pseudocode)

```
1  // Assume item != null and n != null
2  If item < n.data
3      If no left child, insert as n.left
4      Else call recursively on n.left
5  Else if item > n.data
6      If no right child, insert as n.right
7      Else call recursively on n.right
8  Else do nothing (to prevent inserting duplicates)
```

Java Assertions

Notice the comment on line 1 of Listing 9.3. If the public `add()` method has done its job as described above, then neither the item nor the node n are null at the start of `addNode()`. Furthermore, those conditions should continue to be true at the start of any recursive call of `addNode()`. Such a boolean condition that is expected to be true at the start of a method call is known as a **precondition** of the method.

Unfortunately, comments are not executable, so there is nothing preventing the programmer from making a mistake and violating the precondition. A Java **assert** statement essentially makes the comment executable:

```
private void addNode(E item, Node<E> n) {
    assert item != null && n != null;
    ...
}
```

The syntax of an assertion is simple:

```
assert booleanCondition;
```

When assertions are enabled, this statement throws an `AssertionError` if the condition is false; otherwise, it does nothing. Assertions are disabled by default, so to enable them, use the "-ea" or "-enableassertions" flag on the command line or add it to the list of runtime options in your IDE.

Assertions may also be used to check **postconditions**, expected to be true at the end of a method call, or more general **invariants** that are expected to remain true at any particular point in the code.

Exercises

1. Draw the binary search tree that results from inserting each of these sequences of elements into an empty tree:

   (a) 9, 10, 33, 43, 73, 47, 27, 67, 36, 49, 59, 89

   (b) 74, 25, 94, 96, 67, 54, 69, 80, 38, 5, 23, 43

   (c) 69, 49, 59, 21, 61, 82, 44, 91, 36, 1, 95, 74

   (d) 5, 38, 39, 94, 95, 85, 44, 88, 60, 33, 1, 73

   (e) 98, 65, 20, 44, 49, 97, 38, 75, 29, 43, 30, 79

2. Describe the base case(s) for the recursion in Listing 9.3.

3. Implement binary search tree insertion by writing these methods in the `BinarySearchTree` class:

   (a) `add()`   See the discussion on page 157.

   (b) `addNode()`   Implement Listing 9.3. Use an assertion to test the precondition, and remember to set the parent reference of the new node.

4. Add assertions to the following methods in the `BinarySearchTree` class to test appropriate preconditions:

   (a) `predNode()`

   (b) `succNode()`

5. Write a `main()` method for the `BinarySearchTree` class to insert random integers into a binary search tree and print the result.

6. Modify the previous exercise to test each of these queries:

    (a) `contains()`

    (b) `min()`

    (c) `max()`

    (d) `pred()`

    (e) `succ()`

7. Write a `main()` method for the `BinarySearchTree` class to insert several strings into a binary search tree and print the result.

8. Determine the $O()$ performance of binary search tree insertion in terms of h, the height of the tree. Explain your answer.

9.3 Deletion

Deleting from a binary search tree is more complex than other tree operations because it may have to change the internal structure of the tree. For example, insertion is simpler because the new node is always a leaf, which means there is no impact on the rest of the tree.

The first step is to understand how deletion works on paper. Remember that the end result always has to be a valid binary search tree.

Deleting By Hand

The process for deleting a node in a binary search tree depends on the number of children it has.

0 Children Deleting a leaf by hand is easy because it doesn't affect the rest of the tree. For example, deleting 12 in this tree simply removes that node:

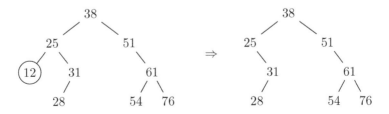

1 Child Deleting a node with one child such as 51 is also not difficult, since we can move its one child 61 up to replace it:

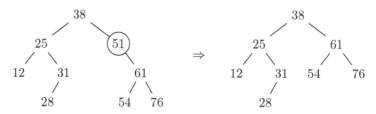

2 Children However, there is no obvious way to delete a node with two children without disrupting the binary search tree. Consider deleting 38 in the tree we just finished with:

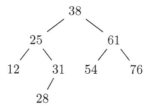

Neither of its children can take 38's place at the root: if 25 moves up, there would be nodes on its left that are too big, and similarly, if 61 moves up, there would be nodes on its right that are too small. (It is also not clear how we would handle the children of 25 or 61.) The solution we adopt is to look elsewhere for a replacement: the inorder predecessor, in this case, 31. It has the right relationship with the other elements in the tree to take the position of 38:

| 12 | 25 | 28 | 31 | 38 | 54 | 61 | 76 |

For the same reason, the inorder successor could also be used.

This approach to deleting a node n with two children involves two steps:

1. Copy the predecessor's value to **n**.
2. Delete the predecessor node.

For example, to delete 38 in this tree, find its predecessor, 31, and copy that value to where 38 was:

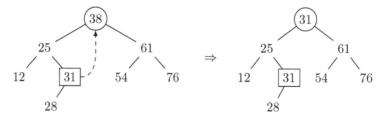

Then delete the predecessor node:

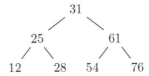

The result is a legal binary search tree without the node 38.

To develop the deletion algorithm, we combine the first two cases into one because they need to do essentially the same work.

Nodes without Two Children

Deleting a node with one child involves replacing that node with its child. But deleting a node with no children can be viewed in the same way: we replace that node with one of its children—it's just that the child happens to be null. Thus, we develop a `replace(n, child)` method that replaces the given node with its child in the tree, and allow for the fact that the child may be null.

There are still three cases to consider, depending on the relationship of n with its parent: it might have no parent (being the root), be a left child, or be a right child:

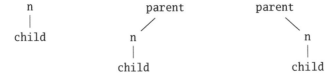

Replacing n with `child` in each of these cases is straightforward. Exercise 5b asks you to implement this `replace()` method.

Putting everything together leads to Listing 9.4, which outlines the private recursive `removeNode()` helper method.

Exercises

1. Suppose a node n has two children in a binary search tree.

    (a) Explain where in the tree the predecessor of n must be located.

    (b) Determine the number of children the predecessor of n has. Explain your answer.

2. Is it always true that the predecessor of a node in a binary search tree has no right child? Explain why or why not.

Listing 9.4: Binary Search Tree Deletion (Pseudocode)

```
1  If n.left is null
2    Replace n with n.right
3  Else if n.right is null
4    Replace n with n.left
5  Else
6    Find predecessor of n
7    Copy data from predecessor to n
8    Recursively delete predecessor
```

3. Draw the binary search tree that results from each of the deletions listed below. In each case, start over with the original tree.

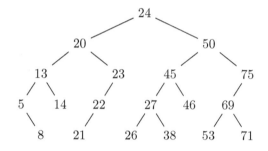

   (a) Delete 45
   (b) Delete 75
   (c) Delete 5
   (d) Delete 20
   (e) Delete 24
   (f) Delete 50

4. For each of these sequences of elements, first draw the binary search tree created by inserting the sequence into an empty tree. Then determine the order in which the elements are removed if the root of the tree is repeatedly deleted until the tree is empty.

   (a) 58, 77, 98, 30, 93, 26, 46
   (b) 64, 38, 6, 36, 57, 62, 40

5. Implement binary search tree deletion in the `BinarySearchTree` class by writing these methods:

   (a) `remove()`   Finds the node to delete and then calls the private `removeNode()` method, returning its (boolean) result.

   (b) `replace()`   Private helper method outlined on page 161. The child may be null, but if it is not null, it will need an updated parent reference.

   (c) `removeNode()`   Implement Listing 9.4, returning true if the deletion is successful.

6. Line 8 of Listing 9.4 suggests using recursion to delete the predecessor.

   (a) Determine the maximum number of times this recursive function could be called (to delete one element).

   (b) Rewrite this step of `removeNode()` to directly remove the predecessor without recursion.

7. Modify Exercise 5 from Section 9.2 to remove the random items, one by one, printing the resulting tree after each deletion. Store the entries in an array as they are inserted so that you have them available to remove.

8. Determine the $O()$ performance of binary search tree deletion in terms of h, the height of the tree. Explain your answer.

9.4   Performance

We began this chapter by expressing the goal of a data structure that provides fast search, insertion, and deletion. To see how well binary search trees perform, it will help to review a few simpler alternatives. The tables below show a single entry for searches because both successful and unsuccessful searches exhibit similar $O()$ performance, but keep in mind that in other circumstances, they may be different.

Unsorted Lists

An unsorted list could be implemented with either an array or linked list. Both display similar worst-case performance characteristics:

	Search	Insert	Delete
Unsorted Array	$O(n)$	$O(1)$	$O(n)$
Unsorted Linked List	$O(n)$	$O(1)$	$O(n)$

Sorted Lists

Sorted lists provide a potential advantage over unsorted lists in searching. This gain is only for arrays, however, and even there, comes at a cost of slower insertion:

	Search	**Insert**	**Delete**
Sorted Array	$O(\log n)$	$O(n)$	$O(n)$
Sorted Linked List	$O(n)$	$O(n)$	$O(n)$

Binary Search Trees

Our analysis of binary search trees to this point has been in terms of h, the height of the tree. Their general performance is:

	Search	**Insert**	**Delete**
Binary Search Tree	$O(h)$	$O(h)$	$O(h)$

Analyzing these operations in terms of the number of nodes in the tree, n, requires translating from h to n. The basic relationship between n and h is:

$$\log n < h \leq n$$

This numeric relationship is expressed visually by the vastly different shapes binary search trees can have. Consider the two extremes when $n = 7$:

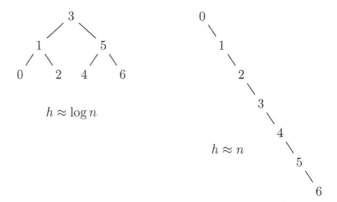

Thus, the worst-case performance for search, insert, and delete in terms of n does not look good:

	Search	**Insert**	**Delete**
Binary Search Tree (Worst Case)	$O(n)$	$O(n)$	$O(n)$

The best case is much better:

	Search	**Insert**	**Delete**
Binary Search Tree (Best Case)	$O(\log n)$	$O(\log n)$	$O(\log n)$

but we should be skeptical of best-case performance most of the time.

Intuitively, most trees will fall somewhere in between, but what will their performance (and shape) look like? What is the shape of an "average" binary search tree? One approach is to consider random elements added to an empty binary search tree. In that case, h is $O(\log n)$ (see Cormen et al. [6] for a proof), and so we can expect that performance with random data:

	Search	**Insert**	**Delete**
Binary Search Tree (Random)	$O(\log n)$	$O(\log n)$	$O(\log n)$

It is important to understand this feature of binary search trees. Their performance is generally fast for all operations, $O(\log n)$, but this bound is *not* guaranteed. All that is guaranteed is $O(h)$, and the worst case is still $O(n)$. **Balanced binary search trees** use various techniques to guarantee that h is $O(\log n)$; see Cormen et al. [6] for the example of **red-black trees**.

Exercises

1. Briefly explain each of the entries for the worst-case performance of unsorted lists.

2. Briefly explain each of the entries for the worst-case performance of sorted lists.

3. Explain why $\log n < h \leq n$ in any binary tree, where h is the height of the tree and n is the number of nodes in the tree.

4. Average-case performance is easier to analyze for unsorted and sorted lists than binary search trees.

   (a) Create a table of the average-case performance of search, insert, and delete for unsorted lists, explaining each entry.

   (b) Create a table of the average-case performance of search, insert, and delete for sorted lists, explaining each entry.

5. Draw all 14 different binary search trees with $n = 4$ elements, and then use them to calculate the average height of a binary search tree of size four.

6. A simple method of allowing duplicate entries in a binary search tree is to change the insertion algorithm in Listing 9.3 to add duplicates in one direction:

```
If data < n.data
    Insert on left
Else
    Insert on right
```

    (a) Modify your code to use this algorithm.

    (b) Describe why this strategy may have a negative impact on performance.

    (c) Outline a different strategy that should reduce this negative impact.

    (d) Implement your strategy.

7. Deleting the predecessor instead of the original node when a node has two children may have a negative impact on performance over time.

    (a) Explain the reason for this effect.

    (b) Outline a different deletion strategy that should reduce this negative impact.

Chapter 10

Heaps

The heap data structure provides fast search, insertion, and deletion by limiting the type of searches and deletions it will do. Like binary search trees, heaps are binary trees, but they allow a looser relationship between elements in order to guarantee their height is always $O(\log n)$.

10.1 Priority Queue Interface and Array-Based Heaps

Priority queues and heaps are closely related because heaps are often used as the underlying data structure for the priority queue interface. However, other structures may be used to implement priority queues, and heaps have purposes beyond just implementing the priority queue interface.

Priority Queue Interface

A **priority queue** is an ordered structure in which items are removed in priority order rather than first-in, first-out like a regular queue. We assume that priorities are assigned by the way in which the items in the queue implement the `Comparable` interface.

The main operations in a priority queue are to insert a new item and to remove the item in the queue with highest priority. Thus, the ADT in Table 10.1 is very similar to that of a regular queue.

If you consider the other data structures we have studied (see Exercise 3), none of them quite fit this interface. Heaps, on the other hand, work nicely.

Heap Data Structure

There are actually two types of heaps, max-heaps and min-heaps, since notions of priority can be based on having either larger or smaller values.

Max-Heaps are complete binary trees (defined shortly) in which every element has value greater than or equal to that of its children.

TABLE 10.1:   Priority Queue ADT

void add(E item) Inserts item into priority queue.
boolean isEmpty() True if priority queue has no elements.
E peek() Returns highest priority item without removing.
E remove() Removes and returns highest priority item.
int size() Number of elements in priority queue.

Min-Heaps are complete binary trees in which every element has value less than or equal to its children.

This requirement that parents have a larger or smaller value than their children is known as the **heap property**.

The most important consequence of the heap property is that the largest element in a max-heap is always at the top of the heap. Similarly, the smallest element in a min-heap is at the top. This makes finding the item with highest priority easy in a heap.

Complete Binary Trees

A **complete binary tree** is a binary tree in which every level except the last has all possible nodes, and on the last level, nodes fill in from the left. For example, the tree on the left is complete but the tree on the right is not:

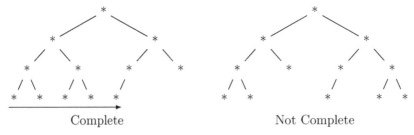

The reason for defining "complete" in this way is that it allows for an efficient array implementation.

Array-Based Binary Trees

General binary trees benefit from the flexibility of a node-based implementation because of the wide variety of shapes they can have. Putting a tight restriction on the tree shape by requiring it to be complete leads to an efficient, array-based alternative to using linked nodes.

The idea is to store the complete tree at the beginning of the array, starting with the root and moving down by levels. For example, in a heap of size 12, there is this correspondence between tree locations and array indices:

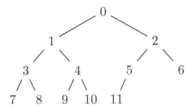

Requiring the tree to be complete is the same as making sure there are no wasted gaps between elements in the array.

Example

Here is an example of a max-heap with ten elements:

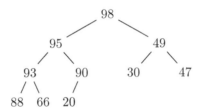

There appears to be less organization than in a binary search tree, because, for example, the left half of the heap has really no relationship with the right half, except that both contain values less than the root. Stored in an array, this heap looks like:

0	1	2	3	4	5	6	7	8	9
98	95	49	93	90	30	47	88	66	20

Parents and Children

To maintain the heap property, we often need to compare the values between parents and their children. Thus, it is helpful to work out the relationship between the array index of a node and those of its parent and children. Using the diagram of indices above, for the node at index i:

$$\text{parent}(i) = (i-1)/2 \quad \text{(using integer division)}$$
$$\text{left}(i) = 2i + 1$$
$$\text{right}(i) = 2i + 2$$

Nicer formulas result if the heap starts at index 1, but that decision complicates other code later.

Generic Arrays of Comparable Elements

The method we have used to create generic arrays does not work if the type parameter E extends the `Comparable` interface. The statement

```
arrayRef = (E[]) new Object[LENGTH];
```

compiles, but if E is declared to extend `Comparable`:

```
<E extends Comparable<? super E>>
```

then the array creation fails at runtime because the `Object` type is not guaranteed to implement `Comparable`.

The solution is to create the array using `Comparable` instead of `Object`:

```
arrayRef = (E[]) new Comparable[LENGTH];
```

An alternative that avoids this problem is to declare the array to have type `Object[]` instead of `E[]`, but then every array access must be cast to type E.

Exercises

1. Show the results of these operations on an initially empty integer priority queue named pq. Give the return value of each non-void operation, and assume that larger values have higher priority. Give the queue contents as a list sorted largest to smallest.

   (a) pq.add(5)
       pq.add(8)
       pq.peek()
       pq.add(3)
       pq.remove()
       pq.add(10)
       pq.size()
       pq.add(4)
       pq.remove()
       pq.remove()

   (b) pq.add(10)
       pq.add(20)
       pq.add(30)
       pq.add(40)
       pq.peek()
       pq.add(50)
       pq.remove()
       pq.size()
       pq.remove()
       pq.remove()

2. Write a generic Java interface `PriorityQueue<E>` for the Priority Queue ADT in Table 10.1.

3. Determine the $O()$ worst-case performance for the `add()` and `remove()` methods in the Priority Queue interface if each of these data structures were to be used for the underlying implementation:

   (a) Unsorted array

   (b) Unsorted linked list

   (c) Sorted array

   (d) Sorted linked list

   (e) Binary search tree

   Explain your answers.

4. Explain why the largest element in a max-heap is at the top.

5. Explain why the smallest element in a min-heap is at the top.

6. Determine the index of the last element in any heap.

7. Explain why the height of a heap is $O(\log n)$.

8. Discuss the difference in memory usage between node-based and array-based trees. Is there an advantage one way or the other? Explain.

9. Determine the parent, left, and right relationships if the heap is stored beginning at index 1 instead of 0.

10. Begin the `MaxHeap<E>` class implementing `PriorityQueue<E>` by declaring instance variables and writing these public methods:

    (a) Constructor(s)   Create arrays using `Comparable` instead of `Object`.

    (b) `size()`

    (c) `isEmpty()`

    (d) `peek()`   Throw a `NoSuchElementException` if the heap is empty.

    (e) `toString()`

    The other public methods will be developed in the next section.

11. Continue the previous exercise by writing these private helper methods:

    (a) `resize()`   Create arrays using `Comparable` instead of `Object`.

    (b) `parent()`, `left()`, and `right()`

    (c) `swap(i, j)`   Swaps elements at indices i and j.

    (d) `isValid(i)`   Returns true if i is a valid heap array index based on the current size of the heap.

    (e) `isLarger(i, j)`   Returns true if the item at index i is strictly greater than the item at index j using `compareTo()`.

12. Repeat Exercises 10 and 11 to begin a `MinHeap<E>` implementation, replacing `isLarger()` with `isSmaller()`.

13. Consider the amount of duplicate code shared between `MaxHeap` and `MinHeap`. Reduce the duplication with an abstract `Heap` class (see Section 8.3).

    Define an **abstract method** `isHigher(i, j)` in the `Heap` class. Abstract methods have no method body; just a declaration that forces any concrete subclass to provide that method. The declaration looks like this:

    ```
    visibility abstract returnType methodName(params);
    ```

    In this case, the concrete subclasses should override `isHigher()` by calling either `isLarger()` or `isSmaller()`.

    The point of this abstract method is that it can be called in `Heap` class methods, representing that item i is higher in the heap than item j, even though it won't be defined until later by concrete subclasses.

10.2   Insertion and Deletion

The key to understanding insertion and deletion in heaps is their shape: because heaps must always be complete binary trees, there is only one slot where a new node can be added and one slot that can be removed.

In the following, we assume heaps are max-heaps; min-heaps work similarly.

Insertion and Heapifying Up

In order to stay complete, a heap must grow at the next open location in the underlying array:

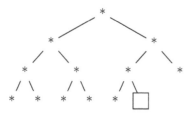

However, this might violate the heap property. If it does, that means the new node is larger than its parent. If we swap the new node with its parent, then any sibling must be ok, and the only possible heap violation is now between the parent and its parent. Continuing this process, called **heapifying up**, works up the tree until we have a legal heap.

Example

Consider inserting 51 into the max-heap below. It begins in the first open space:

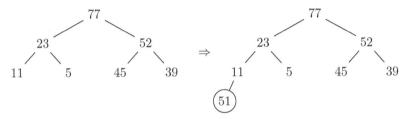

The heap property is violated between 51 and 11, so they swap. At that point, 51 is again larger than its new parent, 23, so they also swap:

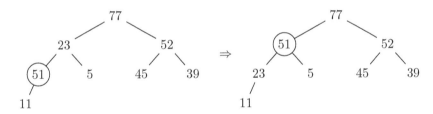

Now 51 is smaller than its parent, and the heap is valid.

Listing 10.1 outlines the algorithm to heapify up from the node i in pseudocode. It is important to check that the parent index is valid so that the algorithm does not attempt to go past the top of the heap.

Listing 10.1: Heapify Up (Pseudocode)

```
1  If parent(i) is valid and i is larger than parent(i)
2    Swap i with parent(i)
3  Repeat
```

Deletion

Deleting from a max-heap means removing and returning the largest element, which is always stored at the root. However, in order to maintain the shape of a complete binary tree, the node that is actually deleted must be at the end of the heap, not the beginning. Therefore, we plan to copy the value from the last node to the root and **heapify down** instead of up.

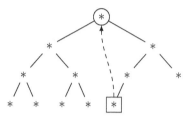

This is similar to the strategy used for deleting in binary search trees: we copy a value from somewhere else and delete that node instead.

One step in the deletion algorithm is easy to miss, and so the process is outlined in Listing 10.2. It is important to change the heap size before heapifying down; otherwise, the last element may participate in the heapification. Line 1 specifies swapping the first and last item instead of just copying the last item for the sake of heapsort in the next section.

Listing 10.2: Heap Deletion (Pseudocode)

```
1  Swap the root with the last item
2  Update the heap size
3  Heapify down
4  Return the largest item
```

Heapifying Down

Heapifying down is more complicated than working up because if there is a violation of the heap property, we can't just look to the one parent—we need to check both children.

Example

Suppose `remove()` is called on the heap we finished with in the last example. Then 77 will be returned, and 11 will need to move to the top (for now, we ignore swapping 77 to the last spot):

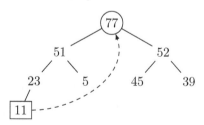

To heapify down with 11 at the top, we need to consider both children and decide which one to swap with 11:

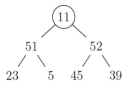

If 11 swaps with 51, then the heap property will be violated between 51 and 52; thus, we swap 11 with its larger child, 52. This process is repeated until we have a valid heap.

Listing 10.3 gives one way to organize the steps in `heapifyDown(i)`. The main task is to decide which of the three items is largest: the value at `i` or one of its two children. This algorithm needs to be careful not to go past the bottom of the heap when accessing children.

Listing 10.3: Heapify Down (Pseudocode)

```
1  Set largest = i
2  If left child is valid and larger than largest, largest = left
3  If right child is valid and larger than largest, largest = right
4  If largest != i
5    Swap i with largest
6    Repeat
```

Exercises

1. Draw the max-heap that results from inserting these sequences of elements into an initially empty heap:

   (a) 37, 17, 81, 74, 95, 52, 8, 83, 41

   (b) 62, 13, 66, 36, 69, 70, 39, 95, 51

   (c) 46, 78, 96, 74, 60, 99, 23, 29, 5

   (d) 40, 60, 57, 82, 55, 4, 6, 97, 59

2. Draw the sequence of max-heaps that result from calling `remove()` three times on the heaps from your answers to Exercise 1.

3. Explain why, when heapifying up, if a node is swapped with its parent, there is no violation of the heap property because of the node's sibling.

4. Implement insertion for max-heaps by writing these methods in the `MaxHeap` class:

    (a) `add()`   Resize if the array is full.

    (b) `heapifyUp()`   Implement Listing 10.1.

5. Implement deletion for max-heaps by writing these methods in the `MaxHeap` class:

    (a) `remove()`   Implement Listing 10.2. Throw an exception if the heap is empty.

    (b) `heapifyDown()`   Implement Listing 10.3.

6. Repeat Exercises 4 and 5 with `MinHeap`.

7. If your `heapifyUp()` method from Exercise 4b is recursive, write a non-recursive version, or vice versa. Is one version preferable?

8. If your `heapifyDown()` method from Exercise 5b is recursive, write a non-recursive version, or vice versa. Is one version preferable?

9. Write a `main()` method for the `MaxHeap` class to insert random integers into a heap and then remove and print each item until the heap is empty.

10. Determine the $O()$ performance of these private heap operations. Explain your answers.

    (a) `heapifyDown()`

    (b) `heapifyUp()`

11. Determine the $O()$ performance of these heap operations. Explain your answers.

    (a) `add()`

    (b) `isEmpty()`

    (c) `peek()`

    (d) `remove()`

    (e) `size()`

12. Determine the $O()$ performance for inserting n items into an empty heap.

10.3   Buildheap and Heapsort

Exercise 12 in the last section explored the cost of inserting n items into an initially empty heap. It turns out there is a faster way to build a heap if all n items are available to begin with.

Buildheap

The buildheap algorithm is based on the observation that if a node is above two heaps:

then calling `heapifyDown()` on the node results in a valid heap rooted at that node. This is essentially what we did in the last section when heapifying down after a deletion.

Buildheap works from the bottom up. Because single nodes are valid heaps, heapifying down from the bottom of the heap can be used to gradually combine nodes together into larger and larger heaps, until the entire tree is a heap.

Example

It is easiest to see how buildheap works with an example. Given the elements 31, 30, 36, 5, 72, 8, 76, 18, 44 in an array, view the array in its heap shape, and focus on the parent of the last item in the heap, along with its children:

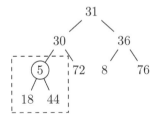

Considering just these three nodes, 5 is a node above two heaps, and so we can heapify down at 5 to create a small, local heap within the larger structure:

Back in context, we then move left one spot in the array to repeat the same process with the next node:

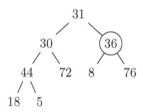

This is also a node above two heaps, and so continuing to heapify down and move left in the array results in this sequence of trees, the last of which is a finished heap:

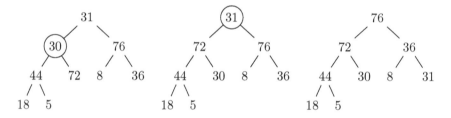

At each stage, because we are working from the bottom up, the two subtrees must already be valid heaps.

Listing 10.4 translates this process into pseudocode.

<hr>

Listing 10.4: Buildheap (Pseudocode)

<hr>

```
1  For each node from the parent of the last node to the root
2     Heapify down
```

<hr>

Buildheap Performance

The buildheap algorithm heapifies $n/2$ times, and each heapify is $O(\log n)$, so an initial estimate of the work done is:

$$O(\frac{n}{2} \log n) = O(n \log n)$$

However, most of the heapifying is done on very short trees near the bottom, and a more careful calculation that takes this into account (see Cormen et al. [6] for details) shows that the worst-case performance of buildheap is $O(n)$.

Heapsort

Buildheap gives a very fast (linear time) way to take an array of items and organize them into a heap. And even though a max-heap puts large elements at the front of the array, it turns out that this organization leads to an efficient sorting algorithm called **heapsort**.

After building the heap, heapsort repeatedly removes the largest item from the heap. Implementing the swap from Line 1 of Listing 10.2 puts the largest item at the end of the array, where it belongs in sorted order.

Example

Consider using heapsort on the array 38, 42, 24, 17, 81, 8, 78. The first step runs buildheap to produce this heap:

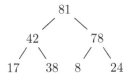

Then the first `remove()` swaps 81 with 24 and calls `heapifyDown()` from the root:

Remember that 81 is no longer in the heap because the size of the heap changed before heapifying down—it stays in the array but not the heap. The next removal swaps 78 and 8, and then heapifies:

This continues until all of the elements have been removed from the heap into their sorted position in the array:

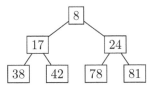

Like buildheap, the pseudocode for heapsort in Listing 10.5 is surprisingly simple.

Listing 10.5: Heapsort (Pseudocode)

```
1  Buildheap
2  While heap is not empty
3      Remove max
```

Exercises

1. Show the sequence of trees and final heap that result from running build-heap on each of these initial arrays:

    (a) 37, 17, 81, 74, 95, 52, 8, 83, 41

    (b) 62, 13, 66, 36, 69, 70, 39, 95, 51

    (c) 46, 78, 96, 74, 60, 99, 23, 29, 5

    (d) 40, 60, 57, 82, 55, 4, 6, 97, 59

2. Show the operation of heapsort on each of these initial arrays. Show the work of buildheap as one step.

    (a) 99, 24, 15, 91, 25, 33, 28

    (b) 7, 65, 29, 10, 39, 59, 79

    (c) 16, 93, 67, 80, 34, 75, 5

    (d) 19, 24, 87, 83, 25, 27, 38

3. Explain why, given a node above two heaps as on page 177, one call to `heapifyDown()` results in a valid heap.

4. Explain in your own words how heapsort produces a sorted list.

5. Determine an expression for the index of the parent of the last node in a heap. You may use the helper functions from Exercise 11 in Section 10.1.

6. Implement a public `MaxHeap(E[] items)` constructor for the `MaxHeap` class that builds a heap from the given array of items. Use a private `buildHeap()` helper method to implement Listing 10.4. Assume the `items` array is full of elements, and keep a default constructor with no parameters for the `MaxHeap` class.

7. Implement a public static `heapSort(T[] items)` method that implements Listing 10.5 to sort the given array of items.

8. Determine the worst-case $O()$ performance of heapsort, and compare it with insertion sort from Section 2.2. Explain your answer.

Project: Event-Based Simulation

The project at the end of Chapter 5 developed a clock-based simulation in which an integer clock variable kept track of the time, ticking along as the program ran. At each time step, the program made decisions about what happened next. This organization is natural but has a limitation: the programmer must decide ahead of time how frequently to tick the clock. The choice is important, because if it doesn't tick often enough, the simulation may miss events, whereas a clock frequency that is too high wastes computation. Using events to organize a simulation instead of a clock eliminates this difficulty.

Events

An **event-based simulation** shifts the focus from the clock to a set of **event objects**. There is still a notion of time, but there is often no central clock. Instead, each event has a time that it happens, and events are processed in order of increasing time. When an event is processed, that time is considered to have "happened," and the event may generate other, later events. Managing these events is easy with a priority queue implemented with a min-heap: events may be added at any time and the next event is always available at the top of the heap.

In an event-based simulation, the clock ticks implicitly when each event is processed from the priority queue, rather than explicitly in a clock variable. In a sense, this clock only ticks when something happens. This eliminates the problem of having to decide how frequently the clock should tick.

Checkout Line Simulation

You may have noticed that some checkout areas with multiple cashiers have a single waiting line that serves all cashiers, while others have separate waiting lines for each. A natural question is whether one of the two methods is more efficient than the other, and one way to approach that question is through simulation.

An event-based checkout simulation might center around Arrival and Departure events. Both types of events store a customer and time, and are ordered by their time. Departure events also store the cashier serving the customer. When an Arrival event is processed, the customer is served by a service system, which is based on either a single queue or multiple queues. The service system either assigns the customer to a cashier or puts the customer in a queue to wait. If there are multiple queues, the customer is put in the shortest line.

When customers are assigned to cashiers, a new Departure event is created based on the service time required by that customer. A Departure event notifies the service system that the cashier is free so that a waiting customer can be assigned to that cashier. The simulation can be initialized by putting a set of Arrival events into the priority queue.

Modeling Customer Arrivals: Poisson Processes

Simulations often employ sophisticated mathematical models for their major components. One of the key components of a checkout line simulation is when customers arrive. A **Poisson process** is often used for this purpose because it assumes no "memory" between arrivals: the time gap before the next customer arrives does not depend on the gap before the previous customer.

If t is the arrival time, the following algorithm computes a sequence of t values for a Poisson process with arrival rate $\lambda > 0$. Begin with $t = 0$, and then for each customer, update t via

$$t = t - \frac{1}{\lambda} \log U$$

where U is a random double between 0 and 1. Continue this process for the desired length of total time. See Ross [11] for details.

Modeling Normal Distributions

Random number generators usually are designed to produce **uniformly distributed** values, meaning that each possible value is equally likely. However, simulations also often require values with an approximately **normal distribution** with a "bell curve" shape. Normal distributions are described by their **mean** and **standard deviation**. See Ross [11] or a statistics textbook for definitions of these terms and more information.

Table 2.3 lists the `nextGaussian()` method of the `java.util.Random` class for generating values from a normal distribution. It assumes a fixed mean of 0.0 and standard deviation 1.0. To use it with different mean `m` and standard deviation `s`, compute:

```
m + gen.nextGaussian() * s
```

where `gen` is a `java.util.Random` object.

Design

Here is a set of classes that may be helpful for implementing this project:

Customer has instance variables for arrival time, service time, and perhaps a unique customer number.

Cashier has an instance variable for the current customer being served.

Server is an abstract class storing an array of cashiers. It defines an

```
assign(customer, cashier, time)
```

method to assign a customer to a particular cashier at a given time. It also defines three abstract methods (see page 172) that are implemented by the concrete subclasses:

```
serve(customer)
clear(cashier, time)
printStats()
```

The subclasses are:

SingleQueueServer uses one queue to hold customers for all cashiers.

MultipleQueueServer has an array of queues for the cashiers.

Event is an abstract class with instance variables for the time and customer. It overrides `compareTo()` so that events compare based on their time. It has one abstract method:

```
process(server)
```

defined by its concrete subclasses:

Arrival events call the server's `serve()` method.

Departure events call the server's `clear()` method.

Simulator is a driver class to run the simulation.

Using this structure, the multiple queue server will need an array of queues that each hold customers. The type of this array will be generic:

```
Queue<Customer>[] queues;
```

If you look ahead to page 193, the `Object` class cannot be used to create this type of generic array; instead the `Queue` type must be used and then cast to the generic type:

```
queues = (Queue<Customer>[]) new Queue[length];
```

Exercises

1. Implement a checkout line simulation as described above to compare the performance of single queue and multiple queue service systems. Use a Poisson process to model customer arrivals and a normal distribution for their required service times. In order to compare the two systems, you will need to submit the same set of customers to each.

2. Add a model of the cashier's efficiency to your simulation, using a normal distribution to produce an efficiency rating for each cashier. Use the efficiency rating as a multiplier on each customer's service time, so that a rating of 1.0 represents an average cashier.

Chapter 11

Hash Tables

Hash tables focus on providing a fast search for any item. They achieve better search performance than even binary search trees by giving up one of the defining features of both binary search trees and heaps: that elements are stored according to their order.

11.1 Map Interface and Linked Implementation

The goal of a hash table is to provide close to $O(1)$ search and insertion for any element. This emphasis on searching leads to a new interface, one based on looking up the **value** of a **key**.

Maps

A **map** is a set of **key-value** pairs in which each key is associated with one value. The same value may be paired with more than one key, but the point is that looking up a key returns the unique value that was stored for that key. Maps are also known as **associative maps**, **associative arrays**, **dictionaries**, or **lookup tables**.

The two main operations of the Map ADT in Table 11.1 are `put()` and `get()`. Deleting from a hash table can be complex, so `remove()` operations are sometimes optional in maps.

The Java convention is to use the type parameter K for key types and V for value types. The key type K should be immutable so that keys do not change after they have been put into a map.

Linked Map

A linked list provides a simple implementation of the map interface that is useful for small sets of key-value pairs. The `get()` and `containsKey()` operations traverse the list until finding the key. The `put()` method also needs to traverse the list to see if there is already an entry for the key, but if not, the new entry can be inserted at the front of the list.

TABLE 11.1:   Map ADT

boolean containsKey(K key) True if key has an entry in map.
V get(K key) Returns value associated with key, null if none.
boolean isEmpty() True if map has no key-value pairs.
V put(K key, V value) Associates key with value in map. If key is already in map, the old value is replaced and returned. Returns null for new entries.
V remove(K key) (Optional) Removes entry for key in map, returning associated value or null if none.
int size() Number of key-value pairs in map.

Implementation of the `LinkedMap<K, V>` class is similar to other linked list code we have written except for one detail: these nodes hold a key-value pair.

Map Entries

Because maps hold key-value pairs, it is useful to create a private nested `Entry<T, U>` class to hold the pairs. Then nodes are declared as:

    Node<Entry<K, V>>

and linked maps are structured like this:

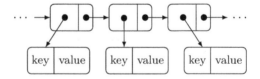

Exercises

1. Are binary search trees a good choice to implement the `Map` interface? Explain why or why not.

2. Write a generic Java interface `Map<K, V>` for the Map ADT in Table 11.1.

3. Modify the `Map` interface from the previous exercise to include the optional `remove()`.

4. Show the results of these operations on an initially empty linked map m. Give the return value of each non-void operation, and draw the final linked list, showing its entry nodes as above.

(a) ```
m.put(5, "five")
m.put(3, "three")
m.containsKey(2)
m.put(5, "FIVE")
m.put(2, "two")
m.get(5)
m.put(1, "one")
m.size()
```

(b) ```
m.put(1, "one")
m.put(2, "two")
m.put(2, "Two")
m.put(2, "TWO")
m.containsKey(2)
m.put(3, "three")
m.get(2)
m.size()
```

5. Develop a `LinkedMap<K, V>` implementation of the `Map` interface based on a linked list. Include these members:

(a) `Entry<T, U>`

(b) `Node<T>`

(c) `containsKey()`

(d) `get()`

(e) `isEmpty()`

(f) `put()`

(g) `size()`

6. Write a `main()` method for the `LinkedMap` class that creates a linked map, puts some values in the map, and looks them up.

7. Determine the $O()$ performance of each of these `LinkedMap` operations:

(a) `containsKey()`

(b) `get()`

(c) `isEmpty()`

(d) `put()`

(e) `size()`

8. Implement the optional `remove()` method in the `LinkedMap` class. The simplest way to do it is by keeping a `prev` pointer to the node before `p` in the traversal.

11.2   Hash Tables

Before describing hash tables, we briefly consider a simpler alternative that can sometimes be useful. In this context, **table** is just another term for an array.

Direct Addressing

The simplest way to get $O(1)$ search and insertion for a map is if keys are relatively small nonnegative integers. Then **direct addressing** can be used, where a table stores values using the integer key as the index:

```
table[key] = value;
```

As long as the table can be allocated large enough to store any possible key, this strategy is simple and efficient.

However, if the keys are not integers or the set of all possible key values is much larger than the desired table size, then a hash function must be used.

Hash Tables

Suppose approximately 2000 records need to be stored, and each record has a four-digit integer key. Then direct addressing is reasonable, because the table will not be too large. However, if each record has a nine-digit integer key, then direct addressing is impractical. The problem is the relationship of the number of possible key values with the desired table size:

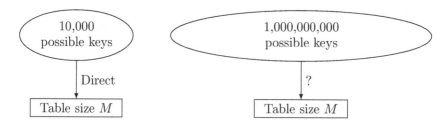

A **hash function** solves this problem by mapping keys to valid table indices:

$$f(\text{key}) = \text{index}$$

A **hash table** is then an array that uses a hash function to compute the indices of its table entries.

Collisions

Of course, if the number of possible key values is much larger than the table size, then **collisions**, in which the hash function maps different keys to the same index, are inevitable.

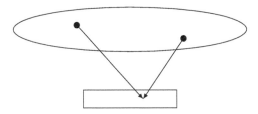

We will look at two **collision resolution strategies:** chaining in Section 11.3 and linear probing in Section 11.4. First we consider how to define hash functions.

Hash Functions

The main job of a hash function is to distribute keys among indices in an approximately uniform way, so that, for example, most keys do not cluster among just a few index values. The name comes from this purpose; it is meant to evoke finely chopped meat and potatoes. A hash function should also be fast to compute and use all aspects of the key.

Integer Keys

In a table of size M, valid array indices are $0, \ldots, M-1$, so the most natural hash function for integer keys is the mod function:

$$f(k) = k \bmod M$$

In Java, the remainder function k % M may be used for modulus as long as k is not negative. This distinction is important for hashing: the Java remainder function can be negative if k itself is negative. (For example, -1 % 3 is -1.) Thus, when using remainder as a hash function, it is important to be sure the value k is nonnegative.

In addition, the table size M should be chosen so that every bit in the key is used. For example, if $M = 2^m$, then only the last m bits of the key are used to compute k % M; see Exercise 2. Because of this, M is often chosen to be prime.

String and Object Keys

Map keys are not always integers, and in fact, are often strings. Each character in a string can be thought of as an integer via its ASCII value. For example, the ASCII values of the characters in the word "scrape" are:

$$
\begin{array}{cccccc}
\text{s} & \text{c} & \text{r} & \text{a} & \text{p} & \text{e} \\
115 & 99 & 114 & 97 & 112 & 101
\end{array}
$$

Adding these values to get a single integer seems reasonable, but it would cause anagrams of "scrape" like "parsec" to hash to the same value. That should be avoided, because a good hash function ought to take into account the order of the letters in the key, not just which letters appear.

The technique used in Java is to interpret the ASCII values as something like a base 31 integer:

$$
\frac{115}{31^5} \quad \frac{99}{31^4} \quad \frac{114}{31^3} \quad \frac{97}{31^2} \quad \frac{112}{31} \quad \frac{101}{1}
$$

$$
= 115 \cdot 31^5 + 99 \cdot 31^4 + 114 \cdot 31^3 + 97 \cdot 31^2 + 112 \cdot 31 + 101
$$

$$
= 3387273908
$$

This method takes into account the order of the characters in the key, although it does produce very large values. The choice of 31 is somewhat arbitrary as a small prime.

The same idea can be used to combine the hash values of multiple fields in an object.

Java Hashcodes

The Java `hashCode()` method implements this algorithm for the `String` type. A minimal `hashCode()` definition is also provided in the `Object` class, with the intention that it be overridden[1] if desired.

Unfortunately, the `hashCode()` method does not always produce valid table indices. For example,

```
"scrape".hashCode()
```

returns -907693388. Why? Because the return type of `hashCode()` is **int**, and the maximum value of an int is $2^{31} - 1$, which is smaller than the hashcode of "scrape" computed above:

$$
2^{31} - 1 = 2{,}147{,}483{,}647 < 3{,}387{,}273{,}908
$$

Therefore, the computation overflows, in this case to a negative value.

---

[1] If you override hashCode(), be sure to adhere to its contract, as specified in the Java API. In particular, if equals() declares that two objects are the same, then they must return the same hashCode().

The result is that Java hashcodes must be forced nonnegative before using them as table indices. A reliable way[2] to do this is to set the first bit to 0:

```
key.hashCode() & Integer.MAX_VALUE
```

The **bitwise & operator** computes the bit-by-bit logical AND of the two values; a single | is used for **bitwise OR**.

Exercises

1. For each of these contexts, decide whether direct addressing or hashing is likely to be more appropriate. Explain your answers.

    (a) About 500 records, each with a 3-digit unique id.

    (b) About 1000 records, each with an 8-digit unique id.

2. Make a table of values of k and k % 8 in both decimal and binary to help explain why using remainder with $M = 2^m$ only uses the last m bits of the key. Choose your own values for k.

3. Give an example of two keys that collide using the hash function

$$f(k) = k \bmod 97$$

4. Give an example of two keys that collide using the hash function

$$f(k) = k \bmod 239$$

5. Calculate the Java hashCode() of each of these strings by hand:

    (a) "Java"
    (b) "table"
    (c) "data"
    (d) "Data"

6. Suppose student registration data will be stored in a hash table, where each registration entry is identified by a numeric student id, a string departmentName, and a numeric courseNumber. Design a hashCode() for the registrations.

7. Suppose a calendar date is stored as three integers: day, month, and year. Design a hashCode() for this date type.

---

[2] Absolute value is the obvious thing to try, but Math.abs() returns a negative value for Integer.MIN_VALUE. Given how **int** values are represented, this problem with absolute value is unavoidable, and so using the more complicated bit operation is a better choice.

8. Explain what exactly happens in this bitwise operation for a 32-bit integer **k**:

   ```
   k & Integer.MAX_VALUE
   ```

9. Explain why `Math.abs()` returns a negative value (see the footnote on page 191). You may need to do some research on integer representations.

11.3   Chaining

Collisions are inevitable when hashing, and one way to resolve collisions is to put all of the items that hash to the same index in a linked list. This method of resolving collisions is called **chaining**, and each linked list is referred to as a **chain**.

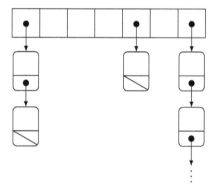

Implementation

A natural way to approach writing a `LinkedHashMap` implementation of a hash table with chaining is to write the linked list code from scratch, as we have before. However, we've already done this work in Section 11.1 by writing the `LinkedMap` class. So a better idea is to just have a linked map at each index of the hash table:

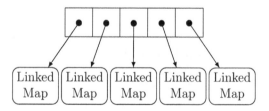

Then each linked map is responsible for the chain at its index.

With this structure, the basic idea for each method in the `LinkedHashMap` class is the same:

1. Use the hash function to compute an index in the hash table.

2. Call the corresponding method on the linked map at that index.

Few complications arise implementing this strategy. The one that may need explanation involves generic arrays.

Generic Arrays Revisited

If each array index stores a reference to a linked map, the table declaration in `LinkedHashMap` will need to be something like this:

```
private LinkedMap<K, V>[] table;
```

As we found with creating arrays for comparable types on page 170, the `Object` class cannot be used to create this generic array. Instead, the `LinkedMap` type must be used and then cast to the generic type:

```
table = (LinkedMap<K, V>[]) new LinkedMap[M];
```

The `Object` type only works for the plain `E[]` arrays we began with:

```
data = (E[]) new Object[capacity];
```

Any more specific (yet generic) type like `LinkedMap<K, V>` requires using the corresponding non-generic `LinkedMap` type to create the array.

Performance

The linear performance of `put()` and `get()` in a linked map may not have seemed very promising in Section 11.1. However, as chains in a hash table, they are impressive because we can control the length of the chains.

The key to determining the performance of a hash table is its **load factor**

$$\lambda = \frac{n}{M}$$

where n is the number of items stored in the hash table and M is the table size. The important fact about the load factor with chaining is this:

If the hash function distributes keys uniformly between 0 and $M - 1$, then the average length of each chain is approximately λ.

This principle leads to the following analysis:

Unsuccessful Search To search for a key that is not in the table, it takes one step to compute the hash function, and then λ steps to check every item in that chain. Thus, the expected number of steps is $O(1 + \lambda)$.

Successful Search A successful search is similar to an unsuccessful one, except that on average,[3] it will only take $\lambda/2$ steps to walk down the chain to find the item. Therefore, in this case, the expected number of steps is $O(1 + \dfrac{\lambda}{2}) = O(1 + \lambda)$.

Insert Insertion in a map requires checking first to see if the key is already present, so the expected performance of insertion is the same as an unsuccessful search, $O(1 + \lambda)$.

Keep in mind that these values are *expected*, based on the ability of the hash function to equally distribute keys. The worst case for each of these operations is $O(n)$.

Space-Time Tradeoff

A hash table is a perfect example of a data structure for which you can buy faster performance with more space: a larger table size M gives a smaller λ and therefore better performance. However, the value of M should be chosen carefully, because if M is much larger than n, most of the array will be wasted storing empty chains.

Deletion

Deleting from a hash table using chaining is routine: it just involves removing a node from the appropriate linked list.

Exercises

1. Show the results of inserting the sequence of keys below into a hash table of size $M = 11$ if collisions are resolved by chaining. Insert new items at the front of each chain.

    (a) 59, 35, 73, 10, 39, 83, 46, 72, 34, 54

    (b) 70, 8, 52, 95, 33, 99, 96, 44, 20, 18

    (c) 28, 32, 13, 78, 86, 98, 20, 57, 68, 83

    (d) 47, 69, 67, 13, 20, 26, 34, 41, 56, 63

---

[3]A careful analysis here is more complex than imagining walking halfway down the chain. See Cormen et al. [6] for details.

2. Show the results of inserting the sequence of keys below into a hash table of size $M = 13$ if collisions are resolved by chaining. Insert new items at the front of each chain.

   (a) 59, 35, 73, 10, 39, 83, 46, 72, 34, 54

   (b) 70, 8, 52, 95, 33, 99, 96, 44, 20, 18

   (c) 28, 32, 13, 78, 86, 98, 20, 57, 68, 83

   (d) 47, 69, 67, 13, 20, 26, 34, 41, 56, 63

3. Is a load factor $\lambda > 1$ possible if chaining is used to resolve collisions? Explain why or why not.

4. Develop a `LinkedHashMap<K, V>` implementation of the `Map` interface that uses chaining to resolve collisions. Use a `LinkedMap` to store each chain. Include these members:

   (a) Constructors Default constructor and a second that takes a table size parameter.

   (b) `hash(key)` Private method to compute the hash function.

   (c) `containsKey()`

   (d) `get()`

   (e) `isEmpty()`

   (f) `put()`

   (g) `size()`

5. Write a `main()` method for the `LinkedHashMap` class that creates a hash table using chains, puts some values in the table, and looks them up.

6. Explain why the worst-case performance for inserting and searching in a hash table using chaining is $O(n)$.

7. Implement the optional `remove()` method in the `LinkedHashMap` class using the `remove()` method from the `LinkedMap` class (Exercise 8 in Section 11.1).

8. Give the expected performance of deleting from a hash table using chaining.

11.4   Linear Probing

An alternative to chaining in a hash table is to store all entries in the array itself. This approach is called **open addressing**. Collisions occur in open addressing when a new element hashes to a location that is already filled in the table. In this case, a **probe sequence** of other table slots is generated, and the new element is stored in the first open slot found by the probe sequence, indicated by a null value in the array:

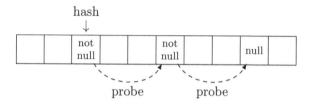

Searching for a key follows the same path: it starts at the hashed location and then uses the same probe sequence until either the key or an open slot is found.

Linear Probing

The probe sequence generated by **linear probing** for a table of size M is given by the function

$$i \rightarrow (i + 1) \bmod M$$

which corresponds to simply moving one slot to the right for each probe, wrapping around to the beginning of the array if necessary.

Example

If the keys 14, 77, and 32 are inserted into a hash table of size 11 the result is (showing primitives for simplicity):

0	1	2	3	4	5	6	7	8	9	10
77			14							32

If the next key is 66, it will also hash to 0 and cause a collision. Using linear probing, the first open slot to the right is 1, and so 66 is stored there:

0	1	2	3	4	5	6	7	8	9	10
77	66		14							32

If the next key is 54, it hashes to 10, and the next open spot for it is index 2:

0	1	2	3	4	5	6	7	8	9	10
77	66	54	14							32

Notice the tendency for keys to **cluster**; this is one of the primary disadvantages of linear probing.

Implementation

As with the `LinkedMap` implementation, a static nested `Entry<K, V>` class is useful for storing key-value pairs in a hash table using open addressing. A separate `probe(i)` function that generates the probe sequence is also helpful to allow easily changing to a different probing method.

Listing 11.1 outlines the method for inserting into a hash table using linear probing. It ignores entries that would force $\lambda \geq 1$.

Listing 11.1: Insertion with Linear Probing (Pseudocode)

```
1  If room in table
2     Hash key
3     While entry not null and not equal key
4        Probe to next entry
5     If entry is null insert new entry
6     Else update old entry
```

Performance

The performance of linear probing is much more complicated to analyze than anything else we have considered to this point. We will simply state the results; interested readers may find the original arguments in Knuth [9]. Sedgewick and Wayne [12] give some intuition for the calculations.

To analyze any open addressing scheme, we need to estimate the number of probes required for a search or insertion. And because we are using open addressing, the load factor λ must be less than one; see Exercises 6 and 7.

Unsuccessful Search An unsuccessful search keeps probing until finding a null table entry. Assuming the hash function distributes keys evenly, the average number of probes for an unsuccessful search is approximately

$$\frac{1}{2}\left(1 + \frac{1}{(1-\lambda)^2}\right)$$

Successful Search A successful search follows the same path that was used to insert the element. However, the element may have been inserted early with a very short path or later with a longer path. In this case, the average number of probes is approximately

$$\frac{1}{2}\left(1 + \frac{1}{1-\lambda}\right)$$

Insertion Inserting an element requires searching for the first open slot in the probe sequence after hashing. This is the same as an unsuccessful search and so the average number of probes is again approximately

$$\frac{1}{2}\left(1 + \frac{1}{(1-\lambda)^2}\right)$$

Deletion

Deleting elements from a table using open addressing is problematic. To see why, consider the example from page 196 created by inserting elements in the order 14, 77, 32, 66, 54:

0	1	2	3	4	5	6	7	8	9	10
77	66	54	14							32

Now suppose 77 is deleted and we search for 54. The search will begin at slot 10, find null in slot 0, and therefore fail. The problem is that items can affect the probe sequence of other elements inserted later than the element being deleted.

There are ways to handle this problem, but none of them are ideal. One option is to use a special **deleted** entry that searches can distinguish from null. Another option is to rehash any elements whose searches would be affected by the deletion (see Listing 11.2 and Exercise 16). However, in most cases, if deletion is important, then a better solution is to use chaining.

Listing 11.2: Deletion with Linear Probing (Pseudocode)

```
1  Find index i containing entry
2  Delete entry
3  Probe to next i
4  While that entry is not null
5     Delete it
6     Rehash it back into the hash table
7     Probe to the next entry
8  Return deleted value
```

Other Probe Sequences

There are several alternatives to linear probing; one of the most useful is known as **double hashing**. The idea in double hashing is to use a second hash function $h(k)$ to determine the step size instead of always taking steps of size 1. This causes different keys to take steps of different sizes.

For double hashing, the probe sequence for a key k is:

$$i \rightarrow (i + h(k)) \bmod M$$

Here it is important for $h(k)$ to be relatively prime to M (see Exercise 13). If M is prime, an easy way to ensure that is to use

$$h(k) = 1 + (k \bmod (M - 1))$$

Exercises

1. Show the results of inserting the sequence of keys below into a hash table of size $M = 11$ if collisions are resolved by linear probing.

   (a) 59, 35, 73, 10, 39, 83, 46, 72, 34, 54

   (b) 70, 8, 52, 95, 33, 99, 96, 44, 20, 18

   (c) 28, 32, 13, 78, 86, 98, 20, 57, 68, 83

   (d) 47, 69, 67, 13, 20, 26, 34, 41, 56, 63

2. Show the results of inserting the sequence of keys below into a hash table of size $M = 13$ if collisions are resolved by linear probing.

   (a) 59, 35, 73, 10, 39, 83, 46, 72, 34, 54

   (b) 70, 8, 52, 95, 33, 99, 96, 44, 20, 18

   (c) 28, 32, 13, 78, 86, 98, 20, 57, 68, 83

   (d) 47, 69, 67, 13, 20, 26, 34, 41, 56, 63

3. Repeat the previous exercise using double hashing with the suggested second hash function $h(k)$.

4. Determine a sequence of key insertions that results in the hash table below if collisions are resolved by linear probing.

0	1	2	3	4	5	6	7	8	9	10
44	77	65	80	14					20	31

5. Calculate the expected number of probes using linear probing for both successful and unsuccessful searches with load factors $\lambda = 0.1$, 0.5, and 0.9.

6. Explain why a load factor $\lambda > 1$ is not possible with open addressing.

7. Explain why a load factor $\lambda = 1$ is not an option with open addressing.

8. What characteristics should a good probe sequence have for open addressing? Explain your answer.

9. Consider the example hash table on page 196 containing the keys 77, 66, 54, 14, and 32. List all the hash values that will cause the existing cluster to grow when inserting the next element.

10. Suppose there is a cluster of size m in a hash table of size M using linear probing. Assuming the hash function distributes keys evenly, determine the probability that the cluster will grow when the next key is added.

11. Develop a `LinearProbeHashMap<K, V>` implementation of the `Map` interface that uses linear probing to resolve collisions. Include these members:

    (a) Constructors   Default constructor and a second that takes a table size parameter.
    (b) `hash(key)`   Private method to compute the hash function.
    (c) `containsKey()`
    (d) `get()`
    (e) `isEmpty()`
    (f) `put()`   Implement Listing 11.1.
    (g) `size()`

12. Write a `main()` method for the `LinearProbeHashMap` class that creates a hash table, puts some values in the table, and looks them up.

13. Explain why the values of the second hash function should be relatively prime to the table size M for double hashing. Hint: try some examples that are not relatively prime to see what happens.

14. Explain why values of the suggested second hash function

$$h(k) = 1 + (k \bmod (M - 1))$$

are guaranteed to be relatively prime to M if M is prime.

15. Array-based implementations of stacks, queues, and lists simply copied their contents into a larger array when their capacity was exceeded.

    (a) Explain why this basic strategy does not work for a hash table using open addressing.
    (b) Describe a strategy that does work.

16. Both chaining and open addressing require extra memory beyond what the entries themselves use. Compare the extra space required for a hash table using chaining with a table using linear probing. Does one method have a clear advantage? Explain your reasoning.

17. Implement the optional `remove()` method in the `LinearProbeHashMap` class. Use the strategy described earlier and outlined in Listing 11.2 of rehashing entries whose searches might be affected by the deletion. Be careful managing the number of entries in the table.

Bibliography

[1] Ken Arnold, James Gosling, and David Holmes. *The Java Programming Language*. The Java Series. Addison-Wesley, Upper Saddle River, NJ, 4th edition, 2006.

[2] Jon Bentley. *Programming Pearls*. ACM Press Books. Addison-Wesley, Reading, MA, 2nd edition, 2000.

[3] Joshua Bloch. Extra, extra - read all about it: Nearly all binary searches and mergesorts are broken. `http://googleresearch.blogspot.com/2006/06/extra-extra-read-all-about-it-nearly.html` (accessed May 15, 2013), June 2006.

[4] Joshua Bloch. *Effective Java*. The Java Series. Addison-Wesley, Upper Saddle River, NJ, 2nd edition, 2008.

[5] Joshua Bloch and Neal Gafter. *Java Puzzlers: Traps, Pitfalls, and Corner Cases*. Addison-Wesley, Upper Saddle River, NJ, 2005.

[6] Thomas H. Cormen, Charles E. Leiserson, Ronald L. Rivest, and Clifford Stein. *Introduction to Algorithms*. The MIT Press, Cambridge, MA, 3rd edition, 2009.

[7] James Cross. jGRASP. `http://www.jgrasp.org/` (accessed May 15, 2013).

[8] James Gosling, Bill Joy, Guy Steele, and Gilad Bracha. *The Java Language Specification*. The Java Series. Addison-Wesley, Upper Saddle River, NJ, 3rd edition, 2005. Available online `http://docs.oracle.com/javase/specs/` (accessed May 15, 2013).

[9] Donald E. Knuth. *The Art of Computer Programming*, volume 3: sorting and searching. Addison-Wesley, Reading, MA, 2nd edition, 1998.

[10] Oracle Corporation. Java API Documentation. `http://docs.oracle.com/javase/7/docs/api/index.html` (accessed May 15, 2013).

[11] Sheldon M. Ross. *Simulation*. Academic Press, San Diego, 3rd edition, 2002.

[12] Robert Sedgewick and Kevin Wayne. *Algorithms*. Pearson Education, Upper Saddle River, NJ, 4th edition, 2011.

[13] Sharon Biocca Zakhour, Sowmya Kannan, and Raymond Gallardo. *The Java Tutorial: A Short Course on the Basics.* The Java Series. Addison-Wesley, Upper Saddle River, NJ, 5th edition, 2013. Available online `http://docs.oracle.com/javase/tutorial/` (accessed May 15, 2013).

Index